T0093697

Introduction to System Reliability Theory

Jorge Navarro

Introduction to System Reliability Theory

 Springer

Jorge Navarro ⓘ
Facultad de Matematicas
Universidad de Murcia
Murcia, Spain

ISBN 978-3-030-86952-6   ISBN 978-3-030-86953-3   (eBook)
https://doi.org/10.1007/978-3-030-86953-3

This Springer imprint is published by the registered company Springer Nature Switzerland AG
The registered company address is: Gewerbestrasse 11, 6330 Cham, Switzerland

To Matilde, Pablo and Julia

Preface

The purpose of this book is to provide the basic tools for a modern post-graduate introductory course on System Reliability Theory. As the excellent classic book by Barlow and Proschan (1975), the present one is just devoted to probabilistic aspects of that theory, including recent results based on signatures, stochastic orders, aging classes, copulas and distortion (or aggregation) functions.

The only requirements for the readers are basic knowledge on Probability Theory and on the Mathematical tools needed in that theory (basic Algebra and Calculus), so that it can serve both for graduate students in Mathematics and for different Engineering students. Some aspects can also be applied to Survival Analysis, Network Reliability or Simple Game Theory. So it could be of interest for other students/researchers in these fields as well.

For that reason, the book includes short introductions to the basic aspects of lifetime modelling, stochastic comparisons, aging classes, mixtures and copula theory needed for the present course. For the interested readers, we provide appropriate references for more advanced results on these topics. Some basic codes written in the statistical (free) program R are also included.

The purpose is to provide the tools for a short basic course (30–60 hours) for different graduate students. So, unfortunately, we have to exclude some relevant aspects on Reliability Theory. For example, the book does not include results based on stochastic processes (since they need more advances courses on probability). Fortunately, there are several recent books on this topic available for the interested readers (see, e.g., Aven and Jensen, 1999; Parzen, 1999; Nakagawa, 2008; Cha and Finkelstein, 2018). Analogously, we do not study here statistical aspects related with reliability data. They are left for a possible second volume.

The book is divided into 5 chapters.

In Chap. 1, we study the basic properties of coherent and semi-coherent binary system structures, obtaining several representations for the structure Boolean function of the system. We do not study non-coherent systems, but we provide some results for multi-state systems with binary components. Relationships with simple games, connectivity properties of networks, mixed systems and fuzzy measures are provided as well.

Chapter 2 is devoted to studying the (random) system lifetime. First, we study its relationship with the component lifetimes. So we introduce the basic tools needed to study random lifetimes as the reliability function, the Mean Time to Failure (MTTF), the hazard (failure) rate function, or the mean residual life function. The different representations based on signatures and distortion functions are studied.

The basic tools to stochastically compare coherent systems are provided in Chap. 3. We study distribution-free comparisons, that is, orderings that do not depend on the components' distribution functions. We consider five cases: systems with Independent and Identically Distributed (IID) components, with Exchangeable (EXC) components, with Identically Distributed (ID) components, with Independent (IND) components and the general case of arbitrary (dependent or independent) components. We use a copula approach to model the dependence structure between the component lifetimes.

In Chap. 4, we study the process of growing old for both the system and the components. To this end, we use the main aging classes (IFR, NBU, DMRL, ILR and their respective dual classes). In particular, we state conditions for the preservation of some of these aging classes under the formation of coherent systems. We also consider different system residual and inactivity times. The limiting behavior (when the time increases) of some system aging functions is studied as well.

In Chap. 5, we study several mechanisms that are used to improve the system's performance. One option is to include some redundant units at some positions in the system. Another popular redundancy option is to add standby components in the system to replace the failed components (when they fail). Another one is to repair these failed components (with perfect or minimal repairs). The main questions analyzed in this chapter are: What is the reliability of the (new) redundant system? What are the best positions to add the redundant components? We also study some component importance indices which can be used to determine the best replacement positions.

I want to thank all the people that helped me in writing this book: my department colleagues, my students and my research collaborators. I do not include names here because I do not know if I will be able to cite all of them.

Murcia, Spain   Jorge Navarro
July 2021

Contents

Acronyms

B	Birnbaum (importance measure)
BFR	Bathtub shaped failure rate
BP	Barlow and Proschan (importance measure)
DD	Diagonal dependent (copula)
DFR	Decreasing failure rate
DFRA	Decreasing failure rate average
DID	Dependent and identically distributed
DLR	Decreasing likelihood ratio
DMIT	Decreasing mean inactivity time
DMRL	Decreasing mean residual life
DRFR	Decreasing reversed failure rate
DTM	Doubly truncated mean
EXC	Exchangeable
FGM	Farlie-Gumbel-Morgenstern (copula)
GPHR	Generalized proportional hazard rate
GPRHR	Generalized proportional reversed hazard rate
HR	Hazard rate
ID	Identically distributed
IFR	Increasing failure rate
IFRA	Increasing failure rate average
IID	Independent and identically distributed
ILR	Increasing likelihood ratio
IMIT	Increasing mean inactivity time
IMRL	Increasing mean residual life
IND	Independent
INID	Independent but not identically distributed
IR	Imperfect repair
IRFR	Increasing reversed failure rate
IT	Inactivity time
LR	Likelihood ratio
LTD	Left tail decreasing
LTI	Left tail increasing
MIT	Mean inactivity time
MR	Minimal repair

MRL	Mean residual life
MTTF	Mean time to failure
NBU	New better than used
NBUE	New better than used in expectations
NQD	Negative quadrant dependent
NWU	New worse than used
NWUE	New worse than used in expectations
ORD	Order
PDF	Probability density function
PHR	Proportional hazard rate
PQD	Positive quadrant dependent
PRHR	Proportional reversed hazard rate
RHR	Reversed hazard rate
RL	Residual life
RR-plot	Reliability-Reliability plot
RTD	Right tail decreasing
RTI	Right tail increasing
SD	Stochastically decreasing
SI	Stochastically increasing
ST	Stochastic (order)

Coherent System Structures

1

Abstract

In this chapter we study the basic properties of the main concept in the Reliability Theory: the coherent system structures. In the first section we give the formal definitions of coherent and semi-coherent (binary) system structures, providing several examples. We do not study non-coherent systems here. We refer the interested readers in that systems to Borgonovo (2010), Imakhlaf et al. (2017) and the references therein. The main properties of coherent systems are given in the second section, including several representations for the structure function of the system. Relationships with simple games, connectivity properties of networks and mixed systems are studied in the third section. The fourth section contains some results for multi-state systems with binary components. The components' importance indices are not studied here. Some of them are studied in Chap. 5. In a first reading, Sects. 1.3 and 1.4 can be skipped (if you want).

1.1 Coherent Structures

The systems are the main concepts in the Reliability Theory. They are "structures" built by using several components. The main assumption is that the state of the system only depends on the states of the components through a "structure function". In this section we assume that the system and the components only have two possible states, a functioning state represented by a 1 and a failure state represented by a 0. Then the formal (mathematical) definition of (binary) *system* can be stated as follows.

Definition 1.1 A (binary) **system** with (binary) components of order n is a Boolean structure function (map)

$$\phi : \{0, 1\}^n \to \{0, 1\},$$

where $\phi(x_1, \ldots, x_n) \in \{0, 1\}$ represents the system's state that is completely determined by the components' states represented by $x_1, \ldots, x_n \in \{0, 1\}$.

To simplify, we just use the word "system" to represent a binary system with binary components. Here it is natural to assume some additional properties for the structure function ϕ. For example, we can expect that a system does not work when all the components fail or that the system works when all the components do so. Analogously, we may also assume that if a broken component is replaced by a functioning component (or it is repaired), then the system state cannot be worse. These assumptions lead to the concept of *semi-coherent systems*. If one (or more) of these properties fails, then we have a non-coherent system that are studied in the references mentioned above.

Definition 1.2 A **semi-coherent system** of order n is a system

$$\phi : \{0, 1\}^n \to \{0, 1\}$$

satisfying the following properties:

 (i) ϕ is increasing;
 (ii) $\phi(0, \ldots, 0) = 0$ and $\phi(1, \ldots, 1) = 1$.

Throughout the book we use the words "increasing" and "decreasing" in a wide sense, that is, a function g is increasing (resp. decreasing) when

$$g(x_1, \ldots, x_n) \leq g(y_1, \ldots, y_n) \quad (\geq)$$

for all $x_1 \leq y_1, \ldots, x_n \leq y_n$.

Semi-coherent systems may have "irrelevant" components, that is, components that do not affect the system. The formal definition is the following.

Definition 1.3 The ith component is **irrelevant** for the system ϕ if

$$\phi(x_1, \ldots, x_{i-1}, 0, x_{i+1}, \ldots, x_n) = \phi(x_1, \ldots, x_{i-1}, 1, x_{i+1}, \ldots, x_n)$$

for all $x_1, \ldots, x_{i-1}, x_{i+1}, \ldots, x_n \in \{0, 1\}$. If this is not the case, then it is a **relevant** component.

For example, the structure function $\phi(x_1, x_2) = x_1$ is a semi-coherent system of order 2 that represents the system formed just with the first component. Here the second component is irrelevant for the system since $\phi(x_1, 0) = \phi(x_1, 1)$ for all x_1. To avoid this problem we consider the concept of coherent system defined as follows. This is the main concept in the present book.

Definition 1.4 A **coherent system** of order n is a system

$$\phi : \{0, 1\}^n \to \{0, 1\}$$

satisfying the following properties:

(i) ϕ is increasing;
(ii) ϕ is strictly increasing in each variable in at least a point.

Clearly, the second condition can be replaced with: "All the components are relevant" and we have the following property.

Proposition 1.1 *All the coherent systems are also semi-coherent systems*

Proof The condition (ii) in the preceding definition implies that, in particular, ϕ is strictly increasing in x_1 in at least a point, that is, there exist $x_2, \ldots, x_n \in \{0, 1\}$ such that

$$0 = \phi(0, x_2, \ldots, x_n) < \phi(1, x_2, \ldots, x_n) = 1.$$

Hence, from (i), we have

$$0 \leq \phi(0, \ldots, 0) \leq \phi(0, x_2, \ldots, x_n) = 0$$

and

$$1 = \phi(1, x_2, \ldots, x_n) \leq \phi(1, \ldots, 1) \leq 1.$$

Therefore, $\phi(0, \ldots, 0) = 0$ and $\phi(1, \ldots, 1) = 1$. $\square$

Note that some semi-coherent systems of order n can be considered as an extension of a coherent system in a dimension $k < n$. For example, the semi-coherent system in dimension 2 defined by $\phi(x_1, x_2) = x_1$ is an extension of the coherent system $\phi(x_1) = x_1$ in dimension 1.

Also note that, from a mathematical point of view, the coherent systems $\phi_1(x_1, x_2, x_3) = \min(x_1, \max(x_2, x_3))$ and $\phi_2(x_1, x_2, x_3) = \min(x_2, \max(x_1, x_3))$ are different. However, when we plot them they have a similar "structure" (see Fig. 1.1). This fact is important when we want to count all the coherent systems of a given dimension (see next section). To consider this fact we need the following definition.

Definition 1.5 We say that two systems ϕ_1 and ϕ_2 of order n are **equivalent under permutations** (shortly written as $\phi_1 \sim \phi_2$) if

$$\phi_1(x_1, \ldots, x_n) = \phi_2(x_{\sigma(1)}, \ldots, x_{\sigma(n)})$$

for a permutation $\sigma : \{1, \ldots, n\} \to \{1, \ldots, n\}$.

Fig. 1.1 Two coherent systems of order 3 with a similar structure

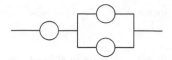

Fig. 1.2 A general structure for a coherent systems of order 3

The equivalence classes determined by this relationship can also be called "systems". For example, the systems given in Fig. 1.1 can be represented by the equivalence class represented by the system in Fig. 1.2.

A coherent (or semi-coherent) system can be determined by the sets of components that assure that the system works (resp. fails) when these components work (fail). The formal definition of such sets is the following.

Definition 1.6 A non-empty set $P \subseteq \{1, \ldots, n\}$ is a **path set** of a system ϕ if $\phi(x_1, \ldots, x_n) = 1$ when $x_i = 1$ for all $i \in P$. A non-empty set $C \subseteq \{1, \ldots, n\}$ is a **cut set** of ϕ if $\phi(x_1, \ldots, x_n) = 0$ when $x_i = 0$ for all $i \in C$. A path set P is a **minimal path set** if it does not contain other path sets. A cut set C is a **minimal cut set** if it does not contain other cut sets.

The sets of path and cut sets of a system ϕ are represented by $\mathcal{P}$ and $\mathcal{C}$. Then we have the following properties. To simplify, in the book, we use "iff" instead of "if and only if".

Proposition 1.2 *Let ϕ be a system. Then:*

(i) ϕ is increasing iff $\mathcal{P}$ is closed under super-inclusions (i.e. if $P \in \mathcal{P}$ and $P \subseteq P^*$, then $P^* \in \mathcal{P}$).
(ii) ϕ is increasing iff $\mathcal{C}$ is closed under super-inclusions.
(iii) $\phi(0, \ldots, 0) = 0$ iff $\{1, \ldots, n\} \in \mathcal{C}$.
(iv) $\phi(1, \ldots, 1) = 1$ iff $\{1, \ldots, n\} \in \mathcal{P}$.
(v) ϕ is semi-coherent iff $\mathcal{P}$ is non-empty, closed under super-inclusions and does not contain the empty set.
(vi) ϕ is semi-coherent iff $\mathcal{C}$ is non-empty, closed under super-inclusions and does not contain the empty set.

Note that, in semi-coherent systems, $\mathcal{P}$ and $\mathcal{C}$ have the same structural properties. To explain this fact we need another concept that can be stated as follows.

Definition 1.7 The **dual system** of a system ϕ is the system

$$\phi^D : \{0, 1\}^n \to \{0, 1\}$$

defined by $\phi^D(x_1, \ldots, x_n) := 1 - \phi(1 - x_1, \ldots, 1 - x_n)$ for all $x_1, \ldots, x_n \in \{0, 1\}$.

The following properties for the dual systems can be proved easily.

Proposition 1.3 *Let ϕ be a coherent (resp. semi-coherent) system and let ϕ^D be its dual system. Then:*

(i) ϕ^D is a coherent (resp. semi-coherent) system.
(ii) A set is a path set of ϕ iff it is a cut set of ϕ^D.
(iii) A set is a cut set of ϕ iff it is a path set of ϕ^D.
(iv) A set is a minimal path set of ϕ iff it is a minimal cut set of ϕ^D.
(v) A set is a minimal cut set of ϕ iff it is a minimal path set of ϕ^D.
(vi) $(\phi^D)^D = \phi$.

Let us see now several examples of coherent and semi-coherent systems. The main structures are series and parallel structures defined as follows.

Definition 1.8 The **series system** of order n is

$$\phi_{1:n}(x_1, \ldots, x_n) := \min(x_1, \ldots, x_n).$$

The **parallel system** of order n is

$$\phi_{n:n}(x_1, \ldots, x_n) := \max(x_1, \ldots, x_n).$$

The **series system** with components in the set P is

$$\phi_P(x_1, \ldots, x_n) := \min_{i \in P} x_i.$$

The **parallel system** with components in the set P is

$$\phi^P(x_1, \ldots, x_n) := \max_{i \in P} x_i.$$

The series and parallel systems of order n are coherent systems but that based on a set P are just semi-coherent systems. Of course, $\phi_{\{1,\ldots,n\}} = \phi_{1:n}$ and $\phi^{\{1,\ldots,n\}} = \phi_{n:n}$ and, in these cases, they are also coherent systems. Moreover, the dual system of ϕ_P is ϕ^P and vice versa.

Note that Boolean functions can be expressed in many different ways. The main options are to use *min* and max operators (as above) or to use polynomials (or multinomials). For example, the series system ϕ_P can also be written as

$$\phi_P(x_1, \ldots, x_n) = \prod_{i \in P} x_i.$$

Note that these options coincide when $x_i \in \{0, 1\}$ but that they are different when we extend these functions to other sets (see next chapter). Analogously, the parallel system ϕ^P can also be written as

$$\phi^P(x_1, \ldots, x_n) = \coprod_{i \in P} x_i$$

where the coproduct $\coprod$ is defined as

$$\coprod_{i \in P} x_i = 1 - \prod_{i \in P} (1 - x_i).$$

For example,

$$\phi_{2:2}(x_1, x_2) = \max(x_1, x_2) = x_1 \amalg x_2 = 1 - (1 - x_1)(1 - x_2) = x_1 + x_2 - x_1 x_2$$

for all $x_1, x_2 \in \{0, 1\}$.

We will see in the next section that all the coherent (or semi-coherent) systems can be written by using series and parallel structures.

Other relevant structures are the *k-out-of-n systems* that work when at least k of their n components work. The explicit definition is the following.

Definition 1.9 The **k-out-of-n system** is defined by

$$\phi_{n-k+1:n}(x_1, \ldots, x_n) = \begin{cases} 1, & \text{if } x_1 + \cdots + x_n \geq k \\ 0, & \text{if } x_1 + \cdots + x_n < k \end{cases} \tag{1.1}$$

for $k = 1, \ldots, n$.

The minimal path sets of the k-out-of-n system are all the sets with exactly k components. So it has $\binom{n}{k}$ minimal path sets. Note that with this definition the 1-out-of-n system is the parallel system $\phi_{n:n}$ and the n-out-of-n system is the series system $\phi_{1:n}$. If $(x_{1:n}, \ldots, x_{n:n})$ represents the increasing ordered vector obtained from $(x_1, \ldots, x_n)$, then

$$\phi_{n-k+1:n}(x_1, \ldots, x_n) = x_{n-k+1:n}$$

for $k = 1, \ldots, n$. This notation is the same as that used to represent the *order statistics*, that is, the ordered data obtained from a sample (see, e.g., Arnold et al. 2008; David and Nagaraja 2003). For example, the 2-out-of-3 system is

$$\phi_{2:3}(x_1, x_2, x_3) = x_{2:3} = \max(\min(x_1, x_2), \min(x_1, x_3), \min(x_2, x_3)).$$

Note that this system cannot be plotted in a plane graph similar to that showed in Fig. 1.1 (we need to repeat the components). An alternative representation as a network will be showed in Sect. 1.3.

Other authors prefer to consider the *k-out-of-n:F systems* (here F means "failed") that fail when at least k of their n components fail. Its structure function is $\phi_{k:n}$ as defined in (1.1) for $k = 1, \ldots, n$. The minimal cut sets of the k-out-of-n:F system are all the sets with exactly k components. So it has $\binom{n}{k}$ minimal cut sets. In this case, the k-out-of-n system considered in the preceding definition can also be called k-out-of-n:G system (here G means "good"). Of course, the dual system of the k-out-of-n:G system is the k-out-of-n:F system and vice versa. Moreover, the k-out-of-n:F system coincides with the $(n - k + 1)$-out-of-n system for $k = 1, \ldots, n$. So we do not need to use the concept of k-out-of-n:F system. However, this notation is needed in the concepts of linear and circular systems defined as follows.

Definition 1.10 For $k = 1, \ldots, n$, the **k-out-of-n:G linear system** is the system that works when at least k consecutive components work, that is, its structure function $\phi_{k:n:G|l}(x_1, \ldots, x_n) = 1$ iff there exists $i \in \{0, \ldots, n-k\}$ such that $x_{i+1} = \cdots = x_{i+k} = 1$. The **$k$-out-of-$n$:$F$ linear system** is the system that fails when at least k consecutive components fail, that is, its structure function $\phi_{k:n:F|l}(x_1, \ldots, x_n) = 0$ iff there exists $i \in \{0, \ldots, n-k\}$ such that $x_{i+1} = \cdots = x_{i+k} = 0$.

The circular systems $\phi_{k:n:G|c}$ and $\phi_{k:n:F|c}$ are defined in a similar way but placing the components in a circle (that is, in this case the first and the last components are also consecutive).

These systems have several applications in practice. For example, the k-out-of-n:F linear systems are used to represent transportation systems as oil or gas pipeline systems and k-out-of-n:F circular systems can represent particle accelerators.

In this case, some k-out-of-n:F linear systems cannot be represented as k-out-of-n:G linear systems. For example, the 2-out-of-3:F linear system is

$$\phi_{2:3:F|l}(x_1, x_2, x_3) = \max(x_2, \min(x_1, x_3)).$$

Its minimal path sets are $P_1 = \{2\}$ and $P_2 = \{1, 3\}$ and its minimal cut sets are $C_1 = \{1, 2\}$ and $C_2 = \{2, 3\}$. So it cannot be represented as a k-out-of-3:G linear system. It is the dual system of the 2-out-of-3:G linear system given by

$$\phi_{2:3:G|l}(x_1, x_2, x_3) = \min(x_2, \max(x_1, x_3)).$$

We conclude this section by computing all the coherent and semi-coherent systems with orders 1-3. Of course, if $n = 1$, then we just have a component and a coherent system $\phi_{1:1}(x_1) = x_1$. If $n = 2$, then we have two coherent systems, the series system $\phi_{1:2}(x_1, x_2) = \min(x_1, x_2)$ and the parallel system $\phi_{2:2}(x_1, x_2) = \max(x_1, x_2)$ of order 2, and the two semi-coherent systems formed with each component. If $n = 3$, then we obtain all the semi-coherent systems given in Table 1.1. Only the nine systems in lines 1, 5, 6, 7, 11, 12, 13, 14 and 18 are coherent systems of order 3. The others are just semi-coherent systems or coherent system of order 1 or 2. The horizontal lines determine the systems that are equivalent under permutations (we have 5 coherent systems and 3 that are just semi-coherent). Note that the system in line $18 - i + 1$ is the dual system of that in line i for $i = 1, \ldots, 7$. The dual systems of the systems in lines 8, 9, 10, 11 are themselves.

1.2 Main Properties

The coherent systems (as Boolean functions) can be written by using different (equivalent) representations. Let us see some of them. The first one is called the *pivotal decomposition* in Barlow and Proschan (1975), p. 5, and can be stated as follows. We shall use the following notation. If $\mathbf{x} = (x_1, \ldots, x_n)$ and $i \in \{1, \ldots, n\}$, then

$$\mathbf{1}_i(\mathbf{x}) = (x_1, \ldots, x_{i-1}, 1, x_{i+1}, \ldots, x_n)$$

Table 1.1 Semi-coherent systems of order 3

N	$\phi_N(x_1, x_2, x_3)$	Minimal path sets	Minimal cut sets
1	$x_{1:3} = \min(x_1, x_2, x_3)$	$\{1, 2, 3\}$	$\{1\}, \{2\}, \{3\}$
2	$\min(x_1, x_2)$	$\{1, 2\}$	$\{1\}, \{2\}$
3	$\min(x_1, x_3)$	$\{1, 3\}$	$\{1\}, \{3\}$
4	$\min(x_2, x_3)$	$\{2, 3\}$	$\{2\}, \{3\}$
5	$\min(x_1, \max(x_2, x_3))$	$\{1, 2\}, \{1, 3\}$	$\{1\}, \{2, 3\}$
6	$\min(x_2, \max(x_1, x_3))$	$\{1, 2\}, \{2, 3\}$	$\{2\}, \{1, 3\}$
7	$\min(x_3, \max(x_1, x_2))$	$\{1, 3\}, \{2, 3\}$	$\{3\}, \{1, 2\}$
8	x_3	$\{3\}$	$\{3\}$
9	x_2	$\{2\}$	$\{2\}$
10	x_1	$\{1\}$	$\{1\}$
11	$x_{2:3}$	$\{1, 2\}, \{1, 3\}, \{2, 3\}$	$\{1, 2\}, \{1, 3\}, \{2, 3\}$
12	$\max(x_3, \min(x_1, x_2))$	$\{3\}, \{1, 2\}$	$\{1, 3\}, \{2, 3\}$
13	$\max(x_2, \min(x_1, x_3))$	$\{2\}, \{1, 3\}$	$\{1, 2\}, \{2, 3\}$
14	$\max(x_1, \min(x_2, x_3))$	$\{1\}, \{2, 3\}$	$\{1, 2\}, \{1, 3\}$
15	$\max(x_2, x_3)$	$\{2\}, \{3\}$	$\{2, 3\}$
16	$\max(x_1, x_3)$	$\{1\}, \{3\}$	$\{1, 3\}$
17	$\max(x_1, x_2)$	$\{1\}, \{2\}$	$\{1, 2\}$
18	$x_{3:3} = \max(x_1, x_2, x_3)$	$\{1\}, \{2\}, \{3\}$	$\{1, 2, 3\}$

and

$$\mathbf{0}_i(\mathbf{x}) = (x_1, \ldots, x_{i-1}, 0, x_{i+1}, \ldots, x_n).$$

Theorem 1.1 (Pivotal decomposition) *Let ϕ be a system of order n, then*

$$\phi(\mathbf{x}) = x_i \phi(\mathbf{1}_i(\mathbf{x})) + (1 - x_i)\phi(\mathbf{0}_i(\mathbf{x})) \tag{1.2}$$

for all $\mathbf{x} = (x_1, \ldots, x_n) \in \{0, 1\}^n$ *and all* $i = 1, \ldots, n$. *Moreover,*

$$\phi(\mathbf{x}) = \sum_{\mathbf{y} \in \{0,1\}^n} \left(\phi(\mathbf{y}) \prod_{j=1}^{n} x_j^{y_j} (1 - x_j)^{1 - y_j} \right) \tag{1.3}$$

for all $\mathbf{x} = (x_1, \ldots, x_n) \in \{0, 1\}^n$.

Proof Clearly, (1.2) holds in the two possible cases, $x_i = 1$ and $x_i = 0$. Expression (1.3) is obtained by repeated applications of (1.2). For example, we can start with x_1 obtaining

$$\phi(\mathbf{x}) = x_1 \phi(1, x_2, \ldots, x_n) + (1 - x_1)\phi(0, x_2, \ldots, x_n).$$

Then we apply (1.2) to $\phi(1, x_2, \ldots, x_n)$ and $\phi(0, x_2, \ldots, x_n)$ for $i = 2$ and so on.   □

Expression (1.3) proves that ϕ can be written as a multinomial of degree n. This representation will be used in the next chapter to compute the reliability of systems with independent components.

For example, the pivotal decomposition for the system

$$\phi_{14}(x_1, x_2, x_3) = \max(x_1, \min(x_2, x_3))$$

(see Table 1.1) is

$$\phi_{14}(x_1, x_2, x_3) = x_1(1 - x_2)(1 - x_3) + (1 - x_1)x_2x_3 + x_1x_2(1 - x_3)$$
$$+ x_1(1 - x_2)x_3 + x_1x_2x_3$$
$$= x_1 + x_2x_3 - x_1x_2x_3.$$

The second representation is based on minimal path or minimal cut sets (defined in the preceding section). It is stated in the following theorem. It will be used in the next chapter to compute the system lifetime and the system reliability.

Theorem 1.2 (Minimal path/cut sets' representations) *Let ϕ be a coherent (or semi-coherent) system of order n and let $P_1, \ldots, P_r$ and $C_1, \ldots, C_s$ be its minimal path and minimal cut sets, respectively. Then*

$$\phi(\mathbf{x}) = \max_{i=1,\ldots,r} \min_{j \in P_i} x_j \tag{1.4}$$

and

$$\phi(\mathbf{x}) = \min_{i=1,\ldots,s} \max_{j \in C_i} x_j \tag{1.5}$$

for all $\mathbf{x} = (x_1, \ldots, x_n) \in \{0, 1\}^n$.

Proof The first expression (1.4) holds since a coherent system works iff at least one of the series systems obtained from its minimal path sets works. Analogously, (1.5) holds since a coherent system fails iff at least one of the parallel systems obtained from its minimal cut sets fails. □

Remark 1.1 The preceding theorem can also be stated by using path or cut sets. However, the expressions obtained in this way are more complicated than that stated above (so we will not use them).

The preceding theorem shows that any coherent system can be decomposed as series systems connected in parallel or as parallel systems connected in series (with some possible common components). Here we can use the notation introduced in the preceding section for series and parallel systems and write (1.4) and (1.5) as

$$\phi(\mathbf{x}) = \max_{i=1,\ldots,r} \phi_{P_i}(\mathbf{x})$$

and

$$\phi(\mathbf{x}) = \min_{i=1,\ldots,s} \phi^{C_i}(\mathbf{x}),$$

respectively. They can also be written by using products and coproducts as

$$\phi(\mathbf{x}) = \coprod_{i=1}^{r} \prod_{j \in P_i} x_j \qquad (1.6)$$

and

$$\phi(\mathbf{x}) = \prod_{i=1}^{s} \coprod_{j \in C_i} x_j. \qquad (1.7)$$

Note that we obtain again multinomials of degre n and that these representations are more "efficient" than the pivotal decomposition. For example, for the system ϕ_{14} considered above, we obtain

$$\phi_{14}(x_1, x_2, x_3) = x_1 \coprod x_2 x_3 = 1 - (1 - x_1)(1 - x_2 x_3) = x_1 + x_2 x_3 - x_1 x_2 x_3.$$

In this chapter, representations (1.4)–(1.7) for the Boolean function ϕ are equivalent. However, in the next chapter, they will be used to extend ϕ to real numbers and then they will provide different expressions (that will be used to different purposes). For example, the series system of order 2 can be written as $\phi_{2:2}(x_1, x_2) = \min(x_1, x_2)$ or as the multinomial $\psi_{2:2}(x_1, x_2) = x_1 x_2$. If $x_1, x_2 \in \{0, 1\}$, then $\phi_{2:2}(x_1, x_2) = \psi_{2:2}(x_1, x_2)$. However, they are different as real functions. For example, $\phi_{2:2}(1/2, 1/2) = 1/2 \neq 1/4 = \psi_{2:2}(1/2, 1/2)$.

The minimal path and minimal cut set representations can also be used to determine all the coherent systems of order n. They show that a system is completely determined by its minimal path sets (or by its minimal cut sets). So a system can also be seen as a finite sequence of subsets of $[n] := \{1, \ldots, n\}$ with the properties given in the following proposition.

Proposition 1.4 *The non-empty sets $P_1, \ldots, P_r \subseteq [n]$ are the minimal path (or cut) sets of a coherent system iff the two following properties hold:*

(i) P_i is not contained in P_j for all $i \neq j$;
(ii) $P_1 \cup \cdots \cup P_r = [n]$.

Proof Clearly, (i) holds when $P_1, \ldots, P_r$ are the minimal path (or cut) sets of a semi-coherent system (by definition). Moreover, if $i \notin P_1 \cup \cdots \cup P_r$ then, from (1.4), the ith component is irrelevant for the system. Therefore, (ii) holds for the minimal path sets of any coherent system. From (1.5), (ii) also holds for the minimal cut sets of a coherent system.

Conversely, if the sets $P_1, \ldots, P_r$ satisfy (i) and (ii), then we can consider the system (Boolean function) ϕ defined by (1.4). Clearly, ϕ is increasing. Moreover, we can prove that any component is relevant due to (ii). Thus, if $i \in [n] = P_1 \cup \cdots \cup P_r$, then, from (ii), there exists a $j \in [r]$ such that $i \in P_j$. Now we consider the point $\mathbf{x} = (x_1, \ldots, x_n)$ defined as $x_k = 1$ if $k \in P_j$ and $x_k = 0$ if $k \notin P_j$. Hence,

$$\phi(\mathbf{0}_i(\mathbf{x})) = 0 < 1 = \phi(\mathbf{1}_i(\mathbf{x}))$$

since $\phi_{P_j}(\mathbf{x}) = x_i$ and $\phi_{P_\ell}(\mathbf{x}) = 0$ for all $\ell \neq j$ (from (i)). Therefore ϕ is a coherent structure. Moreover, it is easy to see that $P_1, \dots, P_r$ are its minimal path sets. The proof for the minimal cut sets is similar.   □

Note that the characteristic properties of minimal path sets and minimal cut sets of coherent systems coincide. This is an expectable property since the minimal path sets of a system are the minimal cut sets of its dual system (and vice versa). However, in the first case we use (1.4) to determine the system while in the second we use (1.5). Moreover, as we have seen in the proof, the minimal path (or cut) sets of semi-coherent systems are just characterized by property (i).

The systems can also be represented by using their paths (or cut) sets. However, as mentioned above, these representations are always more complicated. So we do not include these properties here. Both structures (path/cut sets and minimal path/cut sets) can be used in Set Theory (see Ramamurthy 1990).

The preceding proposition can be used jointly with the following algorithm, extracted from Navarro and Rubio (2010), to determine all the coherent systems of order n. They are determined by their minimal path sets. We use a recursive method on the number k of minimal path sets. We use the notation $|A|$ for the cardinality of the set A. The coherent system ϕ is represented here by the sequence $\phi = (P_1, \dots, P_k)$ of its minimal path sets with $|P_1| \leq \cdots \leq |P_k|$. Some systems can be written in different ways (we avoid repetitions).

Algorithm 1.2.1:
Step 0: Generate the set S with all the non-empty subsets of $[n]$ (there are $m = |S| = 2^n - 1$ subsets).
Step 1: Generate the unique coherent system with $k = 1$ minimal path set (the series system with $P_1 = [n]$). Let $S_1 = \{([n])\}$.
Step 2: Generate all the coherent systems with $k = 2$ minimal path sets by studying (using Proposition 1.4) all the couples of sets from S (there are $\binom{m}{2} = m(m-1)/2$ different couples). Their sequences (P_1, P_2) of minimal path sets are included in the set S_2 with $|P_1| \leq |P_2|$ (avoiding repetitions).
Step k (for k = 3, 4, …): For any sequence $(P_1, P_2, \dots, P_{k-1}) \in S_{k-1}$, generate all the different coherent systems obtained by replacing P_{k-1} with a couple of subsets $A, B \in S$ such that $|P_{k-2}| \leq |A| \leq |B|$. Their sequences of minimal path sets are included in S_k with $|P_1| \leq |P_2| \leq \cdots \leq |P_k|$ (avoiding repetitions).
Final step: Stop when $S_k = \emptyset$.

Theorem 1.3 *The preceding algorithm generates all the coherent systems of order n.*

Proof Clearly, from the preceding algorithm, S_1 and S_2 contain all the coherent systems with $k = 1$ and $k = 2$ minimal path sets.

Let us see that the set S_3 obtained in step 3 contains all the coherent systems with $k = 3$ minimal path sets. Let P_1, P_2, P_3 be the minimal path sets of a coherent system of size n and let us assume that $|P_1| \leq |P_2| \leq |P_3|$. Then we consider two cases:

Case I: If $P_1 \cup P_2 = [n]$, then, from Proposition 1.4, P_1, P_2 are the minimal path sets of a coherent system of size n, that is, $(P_1, P_2) \in \mathcal{S}_2$ or $(P_2, P_1) \in \mathcal{S}_2$. Hence (P_1, P_2, P_3) is generated in step 3 when in (P_1, P_2) we delete P_2 and we add the couple (P_2, P_3) or when in (P_2, P_1) we delete P_1 and we add the couple (P_1, P_3).

Case II: If $P_1 \cup P_2 \neq [n]$, we define $A = [n] - (P_1 \cup P_2)$ and $Q = P_2 \cup A$. Clearly, $P_1 \cup Q = [n]$ and $|P_1| \leq |P_2| < |Q|$. Hence (P_1, Q) are the minimal path sets of a coherent system of size n, with $|P_1| < |Q|$, that is, $(P_1, Q) \in \mathcal{S}_2$ (in that order). Hence (P_1, P_2, P_3) is generated in step 3 when in (P_1, Q) we delete Q and we add the couple (P_2, P_3).

By induction, let us assume that $\mathcal{S}_{k-1}$ contains all the coherent systems with $k-1$ minimal path sets. We want to prove that the same happen for $\mathcal{S}_k$ by using a procedure similar to that used in step $k = 3$. Let ϕ be a coherent system with minimal path sets $P_1, \ldots, P_k$ satisfying $|P_1| \leq \cdots \leq |P_k|$. As above we consider two cases:

Case I: If $P_1 \cup \cdots \cup P_{k-1} = [n]$, then, from Proposition 1.4, $P_1, \ldots, P_{k-1}$ are the minimal path sets of a coherent system of size n, that is, $(P_1, \ldots, P_{k-1}) \in \mathcal{S}_{k-1}$ (in this way or in a permuted version). Hence $(P_1, \ldots, P_k)$ is generated in step 3 when in $(P_1, \ldots, P_{k-1})$ we delete P_{k-1} (or the last set P_j) and we add the couple (P_{k-1}, P_k) (we add the couple (P_j, P_k)).

Case II: If $P_1 \cup \cdots \cup P_{k-1} \neq [n]$, we define $A = [n] - (P_1 \cup \cdots \cup P_{k-1})$ and $Q = P_{k-1} \cup A$. Clearly, $P_1 \cup \cdots \cup P_{k-2} \cup Q = [n]$ and $|P_1| \leq \cdots \leq |P_{k-2}| < |Q|$. Hence $P_1, \ldots, P_{k-2}, Q$ are the minimal path sets of a coherent system of order n, that is, $(P_1, \ldots, P_{k-2}, Q) \in \mathcal{S}_{k-1}$ (in this way or in a permuted version). Moreover, in all these permuted versions, Q is the last set in the sequence since $|P_i| \leq |P_{k-1}| < |Q|$ for $i = 1, \ldots, k-2$. Hence $(P_1, \ldots, P_k)$ is generated in step k when in $(P_1, \ldots, P_{k-2}, Q)$ (or in any of its permuted versions) we delete Q and we add the couple (P_{k-1}, P_k).   □

The preceding theorem can be used to obtain all the coherent systems of order n (we can use a computer to do so). Let us see an example.

Example 1.1 As we have mentioned in the preceding section, there are 9 coherent system of order 3 (see Table 1.1) that are reduced to just 5 coherent system classes of equivalent systems under permutations. They can be obtained by using the preceding algorithm as follows.

Step 0: If $n = 3$, then $\mathcal{S} = \{\{1\}, \{2\}, \{3\}, \{1, 2\}, \{1, 3\}, \{2, 3\}, \{1, 2, 3\}\}$ (with cardinality $2^3 - 1 = 7$) is the set with all the possible minimal path sets.

Step 1: The unique system with $k = 1$ is the series system

$$\phi_1 = (\{1, 2, 3\}) = \min(x_1, x_2, x_3), \quad \mathcal{S}_1 = \{\phi_1\}.$$

Step 2: For $k = 2$, we consider the $\binom{7}{2} = 21$ couples of sets from $\mathcal{S}$, obtaining six coherent systems:

$$\mathcal{S}_2 = \{\phi_{14}, \phi_{13}, \phi_{12}, \phi_5, \phi_6, \phi_7\},$$

where we use the notation of Table 1.1, that is, $\phi_{14} = (\{1\}, \{2, 3\})$, $\phi_{13} = (\{2\}, \{1, 3\})$, $\phi_{12} = (\{3\}, \{1, 2\})$, $\phi_5 = (\{1, 2\}, \{1, 3\})$, $\phi_6 = (\{1, 2\}, \{2, 3\})$, $\phi_7 =$

$(\{1, 3\}, \{2, 3\})\}$. For example, the first one is obtained as follows. First we consider
all the couples that contain the first set $P_1 = \{1\}$. The first option is $(P_1, P_2 = \{2\})$.
It does not determine a proper coherent system since $P_1 \cup P_2 \neq \{1, 2, 3\}$. The
same happen with $(P_1, P_2 = \{3\})$. The next options are $(P_1, P_2 = \{1, 2\})$ and
$(P_1, P_2 = \{1, 3\})$. They do not determine coherent systems since $P_1 \subset P_2$. The next
one is $(P_1, P_2 = \{2, 3\})$ that leads us to system ϕ_{14}.

Step 3: For $k = 3$, we consider the systems in S_2. With the first one $\phi_{14} =$
$(\{1\}, \{2, 3\})$, we delete $\{2, 3\}$ and when we add the couple $(\{2\}, \{3\})$, we obtain
the parallel system $\phi_{18} = (\{1\}, \{2\}, \{3\})$. Analogously, with the fourth $\phi_5 =$
$(\{1, 2\}, \{1, 3\})$, we delete $\{1, 3\}$, and when we add the pair $\{1, 3\}, \{2, 3\}$, we obtain
the 2-out-of-3 system

$$\phi_{2:3} = \phi_{11} = (\{1, 2\}, \{1, 3\}, \{2, 3\}).$$

In the other options we do not obtain new coherent systems.

Step 4: For $k = 4$, we consider the systems in $S_3 = \{\phi_{18}, \phi_{11}\}$. With the first
one $\phi_{18} = (\{1\}, \{2\}, \{3\})$, we delete $\{3\}$ but we cannot obtain coherent systems by
adding $A, B \in S$ with $1 \leq |A| \leq |B|$. The same happen with the second one
$\phi_{11} = (\{1, 2\}, \{1, 3\}, \{2, 3\})$ when we delete $\{2, 3\}$ and we add $A, B \in S$ with
$2 \leq |A| \leq |B|$. Therefore $S_4 = \emptyset$ and so we stop here.   ◀

Shaked and Suárez–Llorens (2003) proved that there are 20 classes of order 4.
Navarro and Rubio (2010) used the preceding theorem to compute the 180 and 16145
classes of coherent systems of order 5 and 6. The systems of order 5 can be seen in
that paper and those with 6 components in:

   https://webs.um.es/jorgenav/miwiki/doku.php?id=coherent_systems.

The last representation is based on the Möbius transform of ϕ. First, we note that
a system ϕ can be seen as a set function

$$\phi : 2^{[n]} \to \{0, 1\},$$

where $2^{[n]}$ represents the set (or class) of all the subsets of $[n]$ and for $J \subseteq [n]$ we
have

$$\phi(J) := \phi(\mathbf{1}_J)$$

and $\mathbf{1}_J := (x_1, \ldots, x_n)$ with $x_i = 1$ if $i \in J$ and $x_i = 0$ if $i \notin J$. Note that the
condition "ϕ is increasing" can be written now as

$$I \subseteq J \Rightarrow \phi(I) \leq \phi(J)$$

(i.e., ϕ is increasing as a set function). Analogously, the conditions $\phi(0, \ldots, 0) = 0$
and $\phi(1, \ldots, 1) = 1$, can be written now as

$$\phi(\emptyset) = 0 \text{ and } \phi([n]) = 1.$$

Hence, a semi-coherent system ϕ can be seen as a normalized (or regular) *fuzzy measure* (see Fantozzi and Spizzichino 2015; Grabisch 2016). In this sense (see, e.g., Grabisch 2016), the **Möbius transform** $\widehat{\phi}$ of ϕ is defined as

$$\widehat{\phi}(I) := \sum_{J \subseteq I} (-1)^{|I|-|J|} \phi(J). \tag{1.8}$$

It satisfies the following property: if $\phi(I) = 0$, then $\widehat{\phi}(I) = 0$ (since $\phi(J) = 0$ for all $J \subseteq I$). Moreover the inverse relation

$$\phi(J) = \sum_{I \subseteq J} \widehat{\phi}(I) \tag{1.9}$$

holds. Thus we obtain the following representation.

Theorem 1.4 (Möbius representation) *The structure function of a coherent system ϕ can be written as*

$$\phi(x_1, \ldots, x_n) = \sum_{I \subseteq [n]} \widehat{\phi}(I) \prod_{i \in I} x_i \tag{1.10}$$

for all $x_1, \ldots, x_n \in \{0, 1\}$, where $\widehat{\phi}$ is Möbius transform of ϕ defined by (1.8).

The proof is immediate from (1.9) taking into account that if $(x_1, \ldots, x_n) = \mathbf{1}_J$, then $I \subseteq J$ iff $\prod_{i \in I} x_i = 1$. The main advantage of this representation is that it gives us directly the coefficients of the multinomial representation (in the other representations, we have to do some calculations). Let us see an example.

Example 1.2 Let us consider again the coherent system
$$\phi_{14}(x_1, x_2, x_3) = \max(x_1, \min(x_2, x_3)).$$
Its Möbius transform is given by

$$\widehat{\phi}_{14}(\{1\}) = \sum_{J \subseteq \{1\}} (-1)^{1-|J|} \phi(J) = (-1)^0 \phi(\{1\}) = 1,$$

$$\widehat{\phi}_{14}(\{2, 3\}) = \sum_{J \subseteq \{2,3\}} (-1)^{2-|J|} \phi(J) = (-1)^0 \phi(\{2, 3\}) = 1,$$

$$\widehat{\phi}_{14}(\{1, 2, 3\}) = \sum_{J \subseteq \{1,2,3\}} (-1)^{3-|J|} \phi(J)$$

$$= (-1)^{3-1} \phi(\{1\}) + (-1)^{3-2} \phi(\{2, 3\}) + (-1)^{3-2} \phi(\{1, 2\})$$

$$+ (-1)^{3-2} \phi(\{1, 3\}) + (-1)^{3-3} \phi(\{1, 2, 3\})$$

$$= -1$$

and $\widehat{\phi}_{14}(I) = 0$ for the other subsets I. Therefore, from (1.10), we obtain

$$\phi_{14}(x_1, x_2, x_3) = x_1 + x_2 x_3 - x_1 x_2 x_3$$

as in the preceding examples.   ◀

For more properties on systems' structures we refer the readers to Barlow and Proschan (1975), Marichal et al. (2011) and Ramamurthy (1990).

1.3 Related Concepts

Coherent system structures are similar to other concepts considered in different mathematical and engineering subjects. Let us see some of them.

1.3.1 Simple Games

Let N be a finite set and let 2^N be its power set (with all the subsets of N). Here the elements of N are called **players** and the elements of 2^N are called **coalitions** (see Ramamurthy 1990, p. 37). Then a **simple game** (or a **voting system**) on N is defined as follows.

Definition 1.11 A **simple game** on N is $\lambda : 2^N \to \{0, 1\}$ such that:

   (i) $\lambda(\emptyset) = 0$;
  (ii) $\lambda(N) = 1$;
 (iii) $\lambda(A) \leq \lambda(B)$ for all $A \subseteq B$.

A coalition A is a **winning (losing) coalition** if $\lambda(A) = 1$ (0). It is a **blocking coalition** if $\lambda(A^c) = 0$ where $A^c = N - A$. A winning (blocking) coalition is minimal if it does not contain other winning (blocking) coalitions. To simplify, we can assume $N = [n]$. A player $i \in N$ is called a **dictator** if $\{i\}$ is winning and it is called a **veto-player** if $\{i\}$ is blocking. A player $i \in N$ is called a **dummy** if $\lambda(\{i\} \cap A) = \lambda(A)$ for all A.

Clearly, simple games are equivalent to semi-coherent systems, replacing players with components, winning coalitions with path sets, blocking coalitions with cut sets and dummy players with irrelevant components.

The axioms (properties) that must satisfy a simple game are the following (see Ramamurthy 1990, p. 37).

A1. Every coalition is either winning or losing.
A2. The empty set is losing.
A3. The all player set N is winning.
A4. No losing coalition contains a winning coalition.

Sometimes, the following axioms are also added:

A5. If A is winning, then A^c is losing (proper games).
A6. If A is losing, then A^c is winning (strong games).

The simple games can be classified (see Ramamurthy 1990, p. 42–43) as follows.

1. **Proper games.** Every winning coalition is also a blocking coalition. In this case N cannot be divided in two disjoint winning coalitions. This prevent to get different decisions from disjoint coalitions.
2. **Strong games.** Every blocking coalition is also a winning coalition. In this case N cannot be divided in two disjoint blocking coalitions. This prevent to get a blocking situation from disjoint coalitions.
3. **Decisive games.** They are both proper and strong games.
4. **Symmetric games.** There exists an integer number k such that A is winning iff $|A| \geq k$. These games are equivalent to k-out-of-n systems.
5. **Weighted majority games.** There exist a non-negative vector of weights $(w_1, \ldots, w_n)$ and a real number r such that A is winning iff $\sum_{i \in A} w_i \geq r$. In particular it is also **homogeneous** if all the minimal winning coalitions have the same weights.

1.3.2 Networks

The networks are everywhere today. There are several problems related with networks. Here we just consider connectivity problems. From a mathematical point of view, they can be defined as follows. The main results of this section have been obtained from Gertsbakh and Shpungin (2010, 2020). These references can also be used to get more results.

Definition 1.12 A **network** is $N = (V, E)$ where V is the vertex (or node) set and E is the edge (or link) set.

Here we just consider networks with a finite set V with $|V| = m$ and a finite set E with $|E| = n$. Usually, the set E is written as $E = \{e_i = \{u_i, v_i\} : u_i, v_i \in V, i = 1, \ldots, n\}$ (undirected networks) or as $E = \{e_i = (u_i, v_i) : u_i, v_i \in V, i = 1, \ldots, n\}$ (directed networks). We assume that the vertices do not fail but that the edges can fail. As in the case of systems, we just consider two possible states for the edges (up and down). A network is **connected** (all connectivity criterion) if all the nodes are connected by a chain of edges. Sometimes, we might fix a set of **terminal** vertices $T \subseteq V$ and just consider connectivity problems between these terminal vertices. All the concepts studied for systems can be translated to these connectivity problems by defining the structure (or state) function of the network

$$\phi : \{0, 1\}^n \to \{0, 1\},$$

where $\phi(x_1, \ldots, x_n) = 1$ (resp. 0) if the network satisfies (does not satisfy) the connectivity conditions when just the edges with $x_i = 1$ work.

For example, the network with $V = \{1, 2, 3\}$ and $E = \{e_1 = \{1, 2\}, e_2 = \{1, 3\}, e_3 = \{2, 3\}\}$ might represent three islands connected with three bridges (or three cities connected by regular lines of airplanes), see Fig. 1.3, left. Then the structure function for the all connectivity criterion is $\phi_{2:3}$, that is, we need at least two working edges. In this case, the minimal path sets are $P_1 = \{e_1, e_2\}$, $P_2 = \{e_1, e_3\}$, and $P_3 = \{e_2, e_3\}$. Remember that this coherent system cannot be plotted as a plane

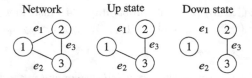

Fig. 1.3 Network of three islands connected with three bridges

system (without repeating components). However, if we just consider the terminal
vertices $T = \{1, 3\}$, then the structure function for the connectivity of these two terminal vertices is $\phi_{13}(x_1, x_2, x_3) = \max(x_2, \min(x_1, x_3))$. In this case, the minimal
path sets are $P_1 = \{e_1, e_3\}$ and $P_2 = \{e_2\}$. For other criterion see Gertsbakh and
Shpungin (2010, 2020).

1.3.3 Mixed Systems

The concept of *mixed system* was introduced by Boland and Samaniego (2004). They
can be used to represent systems that should fulfill different requirements in different
periods of time. They can be defined as follows.

Definition 1.13 We say that ϕ is a **mixed system** of order n if it is equal to ϕ_j with
probability $p_j \geq 0$ for $j = 1, \ldots, m$, where $\phi_1, \ldots, \phi_m$ are systems of order n and
$p_1 + \cdots + p_m = 1$. We say that a mixed system ϕ is **semi-coherent** if $\phi_1, \ldots, \phi_m$
are semi-coherent systems. We say that a mixed system ϕ is **coherent** if $\phi_1, \ldots, \phi_m$
are semi-coherent systems and every component is relevant in at least a system with
a positive probability.

Any (deterministic) system ϕ_1 can be seen as a mixed system ϕ that takes the
value $\phi = \phi_1$ with probability 1. However, the reverse is not true. A mixed system
ϕ can written as a map $\phi : \{0, 1\}^n \rightarrow \{0, 1\}$ but note that here $\phi(x_1, \ldots, x_n)$
represents a discrete random variable that takes the value $\phi_j(x_1, \ldots, x_n) \in \{0, 1\}$
with probability p_j, for $j = 1, \ldots, m$. If ϕ is semi-coherent, then $\phi(0, \ldots, 0) = 0$
and $\phi(1, \ldots, 1) = 1$ (since the same properties hold for any j). However, we cannot
assure that ϕ is increasing (due to the randomness). For example, we can consider
the coherent mixed system ϕ defined as

$$\phi(x_1, x_2, x_3) = \phi_{1:3}(x_1, x_2, x_3) = \min(x_1, x_2, x_3), \text{ with probability } 1/2$$

and

$$\phi(x_1, x_2, x_3) = \phi_{3:3}(x_1, x_2, x_3) = \max(x_1, x_2, x_3), \text{ with probability } 1/2.$$

This mixed system might represent a system (situation) in which we need the three
components half the time (by the day, say) and just one of them in the other half
time (by night). Note that $\phi(0, 0, 0) = 0 \leq \phi(1, 1, 1) = 1$. However, we cannot
assure that $\phi(1, 0, 0) = 0 \leq \phi(1, 1, 0) = 1$ since the following event might happen

$\phi(1, 0, 0) = 1 > \phi(1, 1, 0) = 0$ (with probability $1/4$). Instead we have the following property. If $\mathbf{x} = (x_1, \ldots, x_n)$ and $\mathbf{y} = (y_1, \ldots, y_n)$, we say that $\mathbf{x} \leq \mathbf{y}$ iff $x_i \leq y_i$ for all i.

Proposition 1.5 *If ϕ is a semi-coherent mixed system and $\mathbf{x} \leq \mathbf{y}$, then*

$$E(\phi(\mathbf{x})) \leq E(\phi(\mathbf{y})).$$

Proof From the definition we have

$$E(\phi(\mathbf{x})) = \sum_{j=1}^{m} p_j \phi_j(\mathbf{x}) \leq \sum_{j=1}^{m} p_j \phi_j(\mathbf{y}) = E(\phi(\mathbf{y})),$$

where the inequality holds since $\phi_1, \ldots \phi_m$ are semi-coherent systems. $\square$

1.4 Multi-state Systems with Binary Components

In this section we assume that, for a fixed $m \in \mathbb{N}$, the set of possible states of a system is

$$S := \left\{ 0, \frac{1}{m}, \frac{2}{m}, \ldots, \frac{m-1}{m}, 1 \right\},$$

where, as above, 1 represents the perfect functioning state and 0 the state of failure. In the middle, we have $m - 1$ intermediate states. The evolution in time of the performance of the system can then be seen as a stochastic process starting from 1 (perfect functioning) and eventually going to 0 (failure) as $t \to \infty$.

This representation is clearly equivalent to the classical representation using the levels $\{0, 1, \ldots, m\}$ for a given integer number m. We could of course consider systems with more general levels $\ell_0 = 0 < \ell_1 < \cdots < \ell_m$ by using the set

$$S^* := \left\{ h_0 = 0, h_1 = \frac{\ell_1}{\ell_m}, h_2 = \frac{\ell_2}{\ell_m}, \ldots, h_{m-1} = \frac{\ell_{m-1}}{\ell_m}, h_m = 1 \right\}.$$

This general case can be studied in a similar way.

Thus we define the structure of a multi-state system with binary components as follows.

Definition 1.14 A **multi-state system** with binary components is a function

$$\varphi : \{0, 1\}^n \to S.$$

It is semi-coherent if φ is increasing, $\varphi(0, \ldots, 0) = 0$ and $\varphi(1, \ldots, 1) = 1$. It is coherent if all the components are relevant (i.e. φ is strictly increasing in all the variables in at least a point).

Then we notice that φ has the properties of a normalized (or regular) **fuzzy measure**. As for binary systems, φ can be considered as a set function defined over the family $2^{[n]}$ of all the subsets of $[n]$ where for $J \subseteq [n]$,

$$\varphi(J) := \varphi(1_J)$$

and $1_J := (x_1, \ldots, x_n)$ with $x_j = 1$ for $j \in J$ and $x_j = 0$ for $i \notin J$. In this sense (see, e.g., Grabisch 2016), the Möbius transform $\widehat{\varphi}$ of φ is

$$\widehat{\varphi}(I) := \sum_{J \subseteq I} (-1)^{|I| - |J|} \varphi(J)$$

and it is such that the inverse relation

$$\varphi(J) = \sum_{I \subseteq J} \widehat{\varphi}(I) \tag{1.11}$$

holds. It is also useful for our purposes below to rewrite the previous equation (1.11) in a slightly different form. For $\mathbf{x} \in \{0, 1\}^n$ and $I \subseteq [n]$ such that $\mathbf{x} = 1_I$, we can write

$$\varphi(x_1, \ldots, x_n) = \sum_{J \subseteq I} \widehat{\varphi}(J) = \sum_{J \subseteq [n]} \widehat{\varphi}(J) \prod_{j \in J} x_j. \tag{1.12}$$

This expression is similar to the one obtained for binary systems, see (1.10).

1.4.1 Binary Systems Associated to a Multi-state System

Given a multi-state structure φ, we can consider (see Block and Savits 1982; Marichal et al. 2017) the associated binary systems with the following structures

$$\varphi_i(x_1, \ldots, x_n) = \begin{cases} 1, & \text{if } \varphi(x_1, \ldots, x_n) \geq \frac{i}{m} \\ 0, & \text{if } \varphi(x_1, \ldots, x_n) < \frac{i}{m} \end{cases} \tag{1.13}$$

for $i = 1, \ldots, m$. If φ is semi-coherent, the binary structures $\varphi_1, \ldots, \varphi_m$ are semi-coherent binary systems and satisfy $\varphi_1 \geq \cdots \geq \varphi_m$. Moreover, we have

$$\varphi(x_1, \ldots, x_n) = \frac{1}{m} \sum_{i=1}^{m} \varphi_i(x_1, \ldots, x_n). \tag{1.14}$$

Thus any multi-level system can be associated to a mixed system (see the definition in the preceding subsection) as follows. Note that if ϕ is a mixed system, then $E(\phi)$ is a semi-coherent multi-level system (see the preceding subsection).

Definition 1.15 The mixed system ϕ associated to a multi-level system with structure function φ is the one that is equal to the binary system φ_i with probability $1/m$, for $i = 1, \ldots, m$.

Note that

$$E(\phi(x_1, \ldots, x_n)) = \frac{1}{m} \sum_{i=1}^{m} \varphi_i(x_1, \ldots, x_n) = \varphi(x_1, \ldots, x_n).$$

1.4.2 Multi-state System Associated to a Binary System

Conversely, if ψ is a semi-coherent binary system, then we can define an associated multi-state system. First, we consider the semi-coherent series systems associated to the minimal path sets $P_1, \ldots, P_r$ of ψ defined as

$$\psi_{P_j}(x_1, \ldots, x_n) = \min_{i \in P_j} x_i$$

for $j = 1, \ldots, r$. Then we can use these systems to define the multi-state system with binary components associated to ψ as follows.

Definition 1.16 Let ψ be a semi-coherent system. Then the multi-state system $\tilde{\psi}$: $\{0, 1\}^n \to [0, 1]$ associated to ψ is defined by

$$\tilde{\psi}(x_1, \ldots, x_n) = \frac{1}{r} \sum_{j=1}^{r} \psi_{P_j}(x_1, \ldots, x_n). \tag{1.15}$$

Note that the set of possible states of system $\tilde{\psi}$ is

$$S := \left\{ 0, \frac{1}{r}, \frac{2}{r}, \ldots, \frac{r-1}{r}, 1 \right\}.$$

As above, $\tilde{\psi}$ can be seen as a normalized (or regular) fuzzy measure. The meaning of $\tilde{\psi}$ is clear, it represents the proportion of working minimal path sets in the system (note that the multi-state system $\tilde{\psi}$ could also be defined over the set $\{0, \ldots, r\}$ as the number of working minimal path sets). In particular, $\tilde{\psi} = 1$ means that all the minimal path sets are working and $\tilde{\psi} = 0$ that the system has failed (all the minimal path sets have failed). Therefore, $\tilde{\psi}$ is a risk measure for the system that can be used to describe the *system failure process* from the initial state $\tilde{\psi} = 1$ to the final failure state $\tilde{\psi} = 0$ with intermediate states $(r-1)/r, \ldots, 1/r$. This process could also be used to determine replacement or repair policies in the system.

As in the preceding subsection we can define the associated semi-coherent systems $\psi_j : \{0, 1\} \to \{0, 1\}$ for $j = 1, \ldots, r$, defined by

$$\psi_j(x_1, \ldots, x_n) = 1 \Leftrightarrow \tilde{\psi}(x_1, \ldots, x_n) \geq j/r.$$

Then

$$\tilde{\psi}(x_1, \ldots, x_n) = \frac{1}{r} \sum_{j=1}^{r} \psi_j(x_1, \ldots, x_n)$$

and we can define (as in the preceding section) the mixed system associated to $\tilde{\psi}$.

Problems

1. Prove that if ϕ is a coherent system and $A \subseteq [n]$, then either A is a path set or $A^c := [n] - A$ is a cut set.
2. Compute the structure function of a plain with four engines, two in each wing, that can fly whenever at least an engine is working in each wing.
3. Compute all the coherent systems of order 4 and the number of equivalence classes.
4. Compute all the semi-coherent systems of order 4.
5. Compute the structure functions of all the k-out-of-5 systems.
6. Compute the structure function of a k-out-of-4 linear or circular system. Could some of them be written in a simplified way?
7. Obtain the pivotal decomposition of a coherent system of order 3.
8. Obtain the minimal path set representation of a system of order 4.
9. Obtain the minimal cut set representation of a system of order 4.
10. Obtain the Möbius transform representation of a system of order 4.
11. Prove (1.9).
12. Obtain all the coherent systems of order 5 with 3 minimal path sets.
13. Given a coherent system of order 4, obtain an equivalent network.
14. Given a network, obtain an equivalent coherent system.
15. In a parliament, the parties A, B, C and D have 50, 26, 22 and 11 deputies, respectively. If in a majority decision voting system, they can just vote 'yes' or 'no', obtain the associated simple game (system).
16. Obtain the binary systems associated to a multi-state system.
17. Obtain the multi-state system associated to a binary coherent system.

Coherent System Lifetimes

2

Abstract

In the preceding chapter we have studied systems from a "static" point of view (i.e. at a fixed time value). In the present one, we introduce the time variable (usually represented as t) and we analyse the relationships between the component lifetimes and the system lifetime. In particular, we show how to compute the system reliability function from the component reliability functions (by using different representations). We do the same for the main aging functions (hazard rate, mean residual lifetime, reversed hazard rate, etc.) which allow us to describe the behavior of the system when the time goes on.

2.1 Coherent System Lifetimes

Let us assume from now on that $X_1, \ldots, X_n$ are non-negative random variables on a given probability space $(\Omega, \mathcal{S}, \text{Pr})$ that represent the lifetimes of the components in a system. Hence the system lifetime T can be obtained from the component lifetimes as follows.

Proposition 2.1 *If ψ is a semi-coherent system of order n with minimal path and minimal cut sets $P_1, \ldots, P_r$ and $C_1, \ldots, C_s$, then the system lifetime T can be written as*

$$T = \max_{1 \leq j \leq r} \min_{i \in P_j} X_i \tag{2.1}$$

and

$$T = \min_{1 \leq j \leq s} \max_{i \in C_j} X_i. \tag{2.2}$$

23

The proof is immediate. Let us assume from now on that the above expressions (2.1) and (2.2) (or the expressions (1.4) and (1.5)) are used to extend the structure Boolean function ψ to a real valued function $\psi : \mathbb{R} \to \mathbb{R}$ (we use the same notation). Thus, we can write the lifetime of the system just as $T = \psi(X_1, \ldots, X_n)$. For example, the lifetime of a series system with n components is $T = \min(X_1, \ldots, X_n)$. However note that $T \neq X_1 \cdots X_n$ (i.e. we cannot use the product-coproduct representation of the Boolean structure function to obtain the system lifetime).

As a consequence T is also a non-negative random variable (over the same probability space). Another consequence is that $T = X_I$ for an $I \in [n]$ (but not always the same I, that is, I is also a random variable that can take the values $1, \ldots, n$).

Note that the lifetime of the k-out-of-n system coincides with the order statistic $X_{n-k+1:n}$ from $X_1, \ldots, X_n$. Therefore, the coherent systems contain the order statistics (ordered component lifetimes) as particular cases. Also, as a consequence of the preceding proposition, we have that T is equal to a $X_{J:n}$ for $J \in [n]$, that is, we know that the system is going to fail in one of the ordered points $X_{1:n} \leq \cdots \leq X_{n:n}$. In fact, we will show that, under some assumptions, T can be written as a mixture of the k-out-of-n systems.

We conclude this subsection by noting that the systems can also be studied by using stochastic processes. Thus, for a fixed time $t \geq 0$, we can define the Boolean (or Bernoulli) random variables

$$B_i(t) := 1_{\{X_i > t\}}$$

for $i = 1, \ldots, n$, where $1_A = 1$ (resp. 0) if A is true (false) and $B_i(t) = 1$ (resp. 0) means that the ith component is working (has failed) at time t. Hence the system state at time t is

$$B(t) = \psi(B_1(t), \ldots, B_n(t)) = 1_{\{T > t\}}.$$

Conversely, note that $X_i = \sup\{t : B_i(t) = 1\}$ and $T = \sup\{t : B(t) = 1\}$. Here the system performance is represented by the stochastic process $\{B(t)\}_{t \geq 0}$ where we usually assume $B(0) = 1$ and $B(\infty) = 0$. As mentioned in the preface, we will not use this approach in the present book. The interested reader can go to the references cited there.

2.2 Reliability and Aging Functions

As the system and component lifetimes T and $X_1, \ldots, X_n$ are non-negative random variables, we can consider all the functions used to describe the aging process. Of course we can also use the functions used in the probability theory. The main one is the **system reliability** (or **survival**) function $\bar{F}_T$ defined as

$$\bar{F}_T(t) := \Pr(T > t)$$

for all t. We usually assume $\bar{F}_T(0) = 1$ (the system is working at time $t = 0$). $\bar{F}_T$ is always a decreasing function and satisfies $\lim_{t \to \infty} \bar{F}_T(t) = 0$. The same properties are satisfied by the components' reliability functions defined as

$$\bar{F}_i(t) := \Pr(X_i > t)$$

for all t and $i = 1, \ldots, n$. The respective **distribution** (or unreliability) functions are defined as

$$F_T(t) := \Pr(T \leq t) = 1 - \bar{F}_T(t)$$

and

$$F_i(t) := \Pr(X_i \leq t) = 1 - \bar{F}_i(t)$$

for all t and $i = 1, \ldots, n$. Clearly, $F_T(t)$ and $\bar{F}_T(t)$ represent the probabilities of a working or a broken system, respectively, at time t. So the people usually prefer to use $\bar{F}_T(t)$ instead of $F_T(t)$. Moreover, it is easy to see that, for non-negative random variables, the mean or expected value (lifetime) can be computed as

$$E(T) = \int_0^\infty \bar{F}_T(x)dx. \qquad (2.3)$$

A similar expression holds for the components. In Reliability Theory, this value is also called the **Mean Time To Failure** (MTTF).

The components' reliability functions will be modelled with the most usual models (distributions) for non-negative random variables. Then, as we will see in the following sections, the system reliability will be a function of the components' reliability functions.

The most important model in this field is the **exponential** distribution with reliability function

$$\bar{F}_T(t) = \exp(-t/\mu) \text{ for } t \geq 0,$$

where $\mu > 0$ is the expected value (or MTTF). This model is the unique continuous model which satisfies the following property

$$\Pr(T > x) = \Pr(T - t > x | T > t) \text{ for all } t, x \geq 0.$$

This property is called the **lack of memory property** and means that the reliability in this model is the same for new and used units. So this model plays a central role in the reliability theory representing units which do not have aging. These are considered as good units since the reliability is usually lower for used units (natural or positive aging). In the opposite case, the used units (or the system) have greater reliability functions than the new units (unnatural or negative aging). Note that here "positive" does not mean "good".

A good alternative (more flexible) model is the **Weibull** distribution with reliability function

$$\bar{F}(t) = \exp(-(t/\beta)^\alpha) \text{ for } t \geq 0,$$

where $\alpha, \beta > 0$. The parameter α is called the shape parameter and β is the scale parameter. Note that the Weibull model contains the exponential model (obtained when $\alpha = 1$). It also contains models with natural ($\alpha > 1$) and unnatural ($0 < \alpha < 1$)

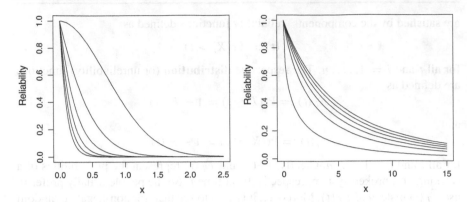

Fig. 2.1 Reliability functions of the residual lifetimes of a Weibull model with $\beta = 1$ and $\alpha = 2$ (left) and $\alpha = 1/2$ (right) for $t = 0, 1, 2, 3, 4, 5$ (from the top on the left and from the bottom on the right)

aging (see below). So it is more flexible than the exponential model. Its distribution function can be computed in R with `pweibull(x,α,β)`.

The random variable $T_t = (T - t | T > x)$ is called the **residual lifetime** (RL) of the system. It represents the performance of used systems that are working at time t. Its reliability function is

$$\bar{F}_T(x|t) := \Pr(T - t > x | T > t) = \frac{\Pr(T > x + t)}{\Pr(T > t)} = \frac{\bar{F}_T(x + t)}{\bar{F}_T(t)} \qquad (2.4)$$

for all $x \geq 0$. It is defined for all $t \geq 0$ such that $\bar{F}(t) > 0$. If $\bar{F}(t) = 0$ for a t, then this random variable does not exist (since the system has already failed for sure at time t). The lack of memory property can also be written as

$$\bar{F}_T(x|t) = \bar{F}_T(x) \text{ for all } t, x \geq 0.$$

The residual lifetime of the system will be studied in Sect. 4.4.

Analogously, the residual lifetimes of the components are $X_{i,t} = (X_i - t | X_i > t)$ and their reliability functions are

$$\bar{F}_i(x|t) := \Pr(X_i - t > x | X_i > t) = \frac{\bar{F}_i(x + t)}{\bar{F}_i(t)}$$

for $i = 1, \ldots, n$ and $t \geq 0$ such that $\bar{F}_i(t) > 0$. They are plotted in Fig. 2.1 for $t = 0, 1, 2, 3, 4, 5$ when the components have a common Weibull reliability with $\beta = 1$ and $\alpha = 2$ (left) and $\alpha = 1/2$ (right). In the left plot they are decreasing in t (positive or natural aging) while in the right plot they are increasing in t (negative aging).

The code for the left plot is the following:

```
R<-function(x) 1-pweibull(x,2,1)
Rt<-function(x,t) R(x+t)/R(t)
curve(R(x),0,2.5,ylab='Reliability')
curve(Rt(x,1),add=T)
curve(Rt(x,2),add=T)
curve(Rt(x,3),add=T)
curve(Rt(x,4),add=T)
curve(Rt(x,5),add=T)
```

There are several functions that can be used to describe the aging process. The first one is the **mean residual life (MRL)** defined as

$$m_T(t) = E(T_t) = E(T - t | T > t)$$

for all $t \geq 0$ such that $\bar{F}(t) > 0$ and that these expectations exist. The MRL functions of the components are defined analogously. From (2.3) and (2.4) it can be computed as

$$m_T(t) = \int_0^\infty \bar{F}_T(x|t)dx = \int_0^\infty \frac{\bar{F}_T(x+t)}{\bar{F}_T(t)}dx = \frac{1}{\bar{F}_T(t)} \int_t^\infty \bar{F}_T(x)dx.$$

Note that it is the area below the residual reliability function $\bar{F}_T(x|t)$ for $t \geq 0$. This function is used to define the **increasing mean residual life** (IMRL) and **decreasing mean residual life** (DMRL) aging classes (according to the monotonicity of m_T). The natural aging is represented by the DMRL class. The exponential model belongs to both classes since its MRL satisfies $m(t) = \mu$ for all $t \geq 0$.

The second one is called the **hazard (or failure) rate (HR or FR)** function and it is defined as

$$h_T(t) = \frac{f_T(t)}{\bar{F}_T(t)}$$

for all t such that $\bar{F}_T(t) > 0$, where $f_T(t) = F_T'(t)$ is a probability density function (PDF) of T (so, note that h_T is not unique). To explain its meaning, we can write it as

$$h_T(t) = \lim_{\epsilon \to 0^+} \frac{\Pr(t < T < t + \epsilon | T > t)}{\epsilon}.$$

Hence, it represents the average probability of failure in the interval $[t, t + \epsilon]$ when $\epsilon \to 0^+$ for a unit that is working at time t.

It is used to define the **increasing failure rate** (IFR) and the **decreasing failure rate** (DFR) aging classes. The exponential model belongs to both classes since its hazard satisfies $h(t) = 1/\mu$ for all $t \geq 0$. In the Weibull model we have $h(t) = \alpha(t/\beta)^{\alpha-1}$ for all $t \geq 0$. Therefore, it is IFR for $\alpha \geq 1$ and DFR for $0 < \alpha \leq 1$.

These aging functions are related (when they exist) by the following expression

$$h(t) = \frac{1 + m'(t)}{m(t)}.$$

So both functions determine the reliability function through the following inversion formula

$$\bar{F}(t) = \exp\left(-\int_0^t h(x)dx\right)$$

for all $t \geq 0$.

Similar properties can be obtained for the **reversed hazard rate** (RHR) and the **mean inactivity time** (MIT) functions. The first one is defined by

$$\bar{h}_T(t) = \frac{f_T(t)}{F_T(t)}$$

for t such that $F_T(t) > 0$ and the second by

$$\bar{m}_T(t) = E(t - T | T \leq t)$$

for all $t \geq 0$ such that these expectations exist. The meaning of $\bar{m}_T$ is clear, it is the expected inactivity time for a system (or unit) that has failed before t. Analogously, $\bar{h}_T(t)$ represents the instantaneous probability of failure at t for a unit that has failed in the interval $[0, t]$. Note that the greater $\bar{h}_T(t)$, the best, since it means that the inactivity time $(t - T | T \leq t)$ is closed to zero. The monotonicity properties of $\bar{h}_T$ and $\bar{m}_T$ are used to define the aging classes IRHR/DRHR and IMIT/DMIT. All these aging classes will be studied in Chap. 4.

2.3 Signature Representations

The first signature representation was obtained by Samaniego (1985) (see also Samaniego 2007). It is based on the fact that the system is going to fail with a component failure. However we need some assumptions. The first one is that the component lifetimes should be independent and identically distributed (IID). In this case, the common distribution (reliability) of the component lifetimes is represented just as F ($\bar{F}$). The second one is that F should be continuous (to avoid ties). Then the representation can be stated as follows.

Theorem 2.1 (Samaniego, 1985) *If T is the lifetime of a coherent system with IID component lifetimes $X_1, \ldots, X_n$ having a common continuous distribution function F, then*

$$\bar{F}_T(t) = \sum_{i=1}^n s_i \bar{F}_{i:n}(t) \tag{2.5}$$

for all t, where $s_1, \ldots, s_n$ are nonnegative coefficients such that $\sum_{i=1}^n s_i = 1$ and that do not depend on F and where $\bar{F}_{i:n}$ is the reliability function of $X_{i:n}$ for $i = 1, \ldots, n$. Moreover, these coefficients satisfy $s_i = \Pr(T = X_{i:n})$ for $i = 1, \ldots, n$.

Proof First note that the events $\{T = X_{i:n}\}$, for $i = 1, 2, \ldots, n$, are a partition of the probability space Ω since, as F is continuous, then $\Pr(X_i = X_j) = 0$ for all $i \neq j$. Hence, from the law of total probability, we have

$$\bar{F}_T(t) = \sum_{i=1}^{n} \Pr(\{T > t\} \cap \{T = X_{i:n}\})$$

$$= \sum_{i=1}^{n} \Pr(T = X_{i:n}) \Pr(T > t | T = X_{i:n})$$

$$= \sum_{i=1}^{n} \Pr(T = X_{i:n}) \Pr(X_{i:n} > t | T = X_{i:n})$$

$$= \sum_{i=1}^{n} \Pr(T = X_{i:n}) \Pr(X_{i:n} > t),$$

where in the sum we only consider the terms with $\Pr(T = X_{i:n}) > 0$ (the others are zero) and where the last equality is obtained from the independence of the events $\{X_{i:n} > t\}$ and $\{T = X_{i:n}\}$ (under the stated assumptions). Thus we obtain (2.5) with $s_i = \Pr(T = X_{i:n})$, for $i = 1, \ldots, n$ and

$$\sum_{i=1}^{n} s_i = \sum_{i=1}^{n} \Pr(T = X_{i:n}) = \Pr(\Omega) = 1$$

which concludes the proof. $\qquad\square$

The vector $\mathbf{s} = (s_1, \ldots, s_n)$ with the coefficients in (2.5) is called the **signature** of the system in Samaniego (1985) (see also Samaniego 2007). It is also called the **destruction spectrum** (or simply **D-spectrum**) when we use networks instead of systems (see, e.g., Gertsbakh and Shpungin 2010, p. 85).

Moreover, these coefficients only depend on the structure of the system (under these assumptions). Actually, they can be computed from ψ as

$$s_i = \frac{|A_i|}{n!} \text{ for } i = 1, \ldots, n, \qquad (2.6)$$

where $|A_i|$ is the cardinality of the set A_i of all the permutations σ of the set $[n] = \{1, \ldots, n\}$ which satisfy that $\psi(x_1, \ldots, x_n) = x_{i:n}$ whenever $x_{\sigma(1)} < \cdots < x_{\sigma(n)}$ (see Samaniego 2007, Chap. 3).

The signature coefficients are also determined by the Boolean structure function ψ as follows

$$s_i = \frac{1}{\binom{n}{i-1}} \sum_{\sum_{j=1}^{n} x_j = n-i+1} \psi(x_1, \ldots x_n) - \frac{1}{\binom{n}{i}} \sum_{\sum_{j=1}^{n} x_j = n-i} \psi(x_1, \ldots x_n) \qquad (2.7)$$

for $i = 1, \ldots, n$ (see Boland 2001).

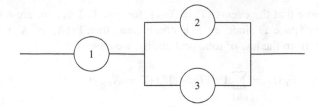

Fig. 2.2 Example of a coherent system

Example 2.1 For the coherent system with lifetime $\psi = \min(x_1, \max(x_2, x_3))$ (see Fig. 2.2), we have the following options (permutations):

σ	$x_{\sigma(1)} < x_{\sigma(2)} < x_{\sigma(3)}$	ψ	J
$(1, 2, 3)$	$x_1 < x_2 < x_3$	$x_1 = x_{1:3}$	1
$(1, 3, 2)$	$x_1 < x_3 < x_2$	$x_1 = x_{1:3}$	1
$(2, 1, 3)$	$x_2 < x_1 < x_3$	$x_1 = x_{2:3}$	2
$(2, 3, 1)$	$x_2 < x_3 < x_1$	$x_3 = x_{2:3}$	2
$(3, 1, 2)$	$x_3 < x_1 < x_2$	$x_1 = x_{2:3}$	2
$(3, 2, 1)$	$x_3 < x_2 < x_1$	$x_2 = x_{2:3}$	2

and hence its signature is $\mathbf{s} = (1/3, 2/3, 0)$. Therefore, from (2.5), its reliability function can be written as

$$\bar{F}_T(t) = \frac{1}{3}\bar{F}_{1:3}(t) + \frac{2}{3}\bar{F}_{2:3}(t)$$

for all t. The signature can also be computed from (2.7) as

$$s_1 = \frac{1}{\binom{3}{0}} \sum_{x_1+x_2+x_3=3} \psi(x_1, x_2, x_3) - \frac{1}{\binom{3}{1}} \sum_{x_1+x_2+x_3=2} \psi(x_1, x_2, x_3) = 1 - \frac{2}{3} = \frac{1}{3},$$

$$s_2 = \frac{1}{\binom{3}{1}} \sum_{x_1+x_2+x_3=2} \psi(x_1, x_2, x_3) - \frac{1}{\binom{3}{2}} \sum_{x_1+x_2+x_3=1} \psi(x_1, x_2, x_3) = \frac{2}{3}$$

and $s_3 = 1 - s_1 - s_2 = 0$.　　　　　　　　　　　　　　　　　　◀

Note that the signature contains the probabilities of the discrete random variable J such that $T = X_{J:n}$. So we can say that T is a **mixture** of $X_{1:n}, \ldots, X_{n:n}$ with weights $s_1, \ldots, s_n$.

If $X_1, \ldots, X_n$ are IID, the ordered variables $X_{1:n}, \ldots, X_{n:n}$ are known as the **order statistics**. In Reliability Theory, they represent the lifetimes of k-out-of-n systems. Their basic properties can be seen in Arnold et al. (2008) and David and Nagaraja (2003). In particular, the expression for their reliability functions are the following (see, e.g., David and Nagaraja 2003, p. 46).

Proposition 2.2 *If $X_1, \ldots, X_n$ are IID~F, then the reliability function of $X_{i:n}$ is*

$$\bar{F}_{i:n}(t) = \sum_{j=0}^{i-1} \binom{n}{j} F^j(t)\bar{F}^{n-j}(t). \tag{2.8}$$

Proof Let us consider the Bernoulli random variables defined by $B_i(t) = 1$ iff $X_i > t$. Then $N(t) := \sum_{i=1}^{n} B_i(t)$ gives the number of components alive at time t. From the IID assumption, $N(t)$ has a Binomial distribution $\mathcal{B}(n, p_t)$ with probability $p_t = \bar{F}(t)$. Therefore

$$\bar{F}_{i:n}(t) = \Pr(X_{i:n} > t) = \Pr(N(t) > n - i) = \sum_{k=n-i+1}^{n} \binom{n}{k} F^{n-k}(t)\bar{F}^k(t)$$

and by doing the change $j = n - k$ we obtain (2.8). $\square$

Note that we can use the expression (2.8) in (2.5) to obtain $\bar{F}_T$ as

$$\bar{F}_T(t) = \sum_{i=1}^{n} s_i \sum_{j=0}^{i-1} \binom{n}{j} F^j(t)\bar{F}^{n-j}(t). \tag{2.9}$$

By interchanging the order of summations in (2.9) we obtain

$$\bar{F}_T(t) = \sum_{k=1}^{n} \left(\sum_{i=n-k+1}^{n} s_i \right) \binom{n}{k} F^{n-k}(t)\bar{F}^k(t), \tag{2.10}$$

where $S_{n-k+1} = \sum_{i=n-k+1}^{n} s_i$ is the probability that the system works when exactly k components work, where $F^{n-k}(t)\bar{F}^k(t)$ is the probability of have k specific components working at age t and where $\binom{n}{k}$ represents the number of options of choosing such k components. An extension of formula (2.10) to the case of non-ID components was obtained in Coolen and Coolen-Maturi (2012) (see also Samaniego and Navarro 2016).

Example 2.2 For the coherent system considered in Example 2.1 with signature $(1/3, 2/3, 0)$, we need

$$\bar{F}_{1:3}(t) = \Pr(X_{1:3} > t) = \bar{F}^3(t)$$

and

$$\bar{F}_{2:3}(t) = \Pr(X_{2:3} > t) = \bar{F}^3(t) + 3\,F(t)\bar{F}^2(t).$$

By replacing F with $1 - \bar{F}$ we get

$$\bar{F}_{2:3}(t) = 3\bar{F}^2(t) - 2\bar{F}^3(t).$$

Note that we do not need

$$\bar{F}_{3:3}(t) = 3\bar{F}(t) - 3\bar{F}^2(t) + \bar{F}^3(t)$$

(since $s_3 = 0$). Hence

$$\bar{F}_T(t) = \frac{1}{3}\bar{F}^3(t) + \frac{2}{3}(3\bar{F}^2(t) - 2\bar{F}^3(t)) = 2\bar{F}^2(t) - \bar{F}^3(t).$$

These reliability functions are plotted in Fig. 2.3, left, when the components are IID with a standard ($\mu = 1$) exponential distribution. The dashed line is the common reliability function of the components. The code in R to get these plots is the following:

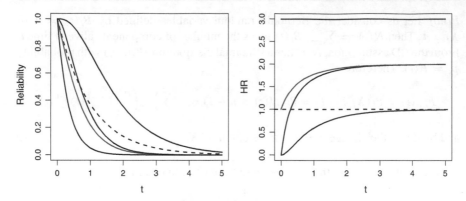

Fig. 2.3 Reliability (left) and hazard rate functions (right) of the system in Example 2.2 (red) and the associated k-out-of-3 systems (black) when the components are IID with a standard exponential distribution. The dashed lines are the common functions for the components

```
R<-function(t) exp(-t)
R1<-function(t) (R(t))^3
R2<-function(t) 3*(R(t))^2-2*(R(t))^3
R3<-function(t) 3*R(t)-3*(R(t))^2+(R(t))^3
s1<-1/3
s2<-2/3
s3<-0
RT<-function(t) s1*R1(t)+s2*R2(t)+s3*R3(t)
curve(RT(x),0,5,xlab='t',ylab='Reliability',col='red',lwd=2)
curve(R1(x),add=T,lwd=2)
curve(R2(x),add=T,lwd=2)
curve(R3(x),add=T,lwd=2)
curve(R(x),add=T,lty=2,lwd=2)
```

Note that

$$\bar{F}_{1:3} \leq \bar{F}_T \leq \bar{F}_{2:3} \leq \bar{F}_{3:3}.$$

This is a general property of this system for all F (due to $s_3 = 0$). Also note that $\bar{F}_T \leq \bar{F}$ but that $\bar{F}$ (dashed line) and $\bar{F}_{2:3}$ (black line in the middle) are not ordered. By changing the signature we can plot other systems.   ◄

   Proceeding as in the preceding example, we can compute the signature vectors for all the coherent systems with 1-4 components. They were first computed by Shaked and Suárez–Llorens (2003) and they are given Table 2.1. Of course, the systems equivalent under permutations have the same signatures (so we just include one of them in the table). However, there are systems not equivalent under permutations that have the same signature as well (see, e.g., the systems with numbers 20 and 21). As a consequence of (2.5), then they also have the same reliability (distribution)

Table 2.1 Signatures of all the coherent systems with 1-4 IID components

i	T_i	s
1	$X_{1:1} = X_1$	(1)
2	$X_{1:2} = \min(X_1, X_2)$ (2-series)	$(1, 0)$
3	$X_{2:2} = \max(X_1, X_2)$ (2-parallel)	$(0, 1)$
4	$X_{1:3} = \min(X_1, X_2, X_3)$ (3-series)	$(1, 0, 0)$
5	$\min(X_1, \max(X_2, X_3))$	$(\frac{1}{3}, \frac{2}{3}, 0)$
6	$X_{2:3}$ (2-out-of-3)	$(0, 1, 0)$
7	$\max(X_1, \min(X_2, X_3))$	$(0, \frac{2}{3}, \frac{1}{3})$
8	$X_{3:3} = \max(X_1, X_2, X_3)$ (3-parallel)	$(0, 0, 1)$
9	$X_{1:4} = \min(X_1, X_2, X_3, X_4)$ (4-series)	$(1, 0, 0, 0)$
10	$\max(\min(X_1, X_2, X_3), \min(X_2, X_3, X_4))$	$(\frac{1}{2}, \frac{1}{2}, 0, 0)$
11	$\min(X_{2:3}, X_4)$	$(\frac{1}{4}, \frac{3}{4}, 0, 0)$
12	$\min(X_1, \max(X_2, X_3), \max(X_3, X_4))$	$(\frac{1}{4}, \frac{7}{12}, \frac{1}{6}, 0)$
13	$\min(X_1, \max(X_2, X_3, X_4))$	$(\frac{1}{4}, \frac{1}{4}, \frac{1}{2}, 0)$
14	$X_{2:4}$ (3-out-of-4)	$(0, 1, 0, 0)$
15	$\max(\min(X_1, X_2), \min(X_1, X_3, X_4),$ $\min(X_2, X_3, X_4))$	$(0, \frac{5}{6}, \frac{1}{6}, 0)$
16	$\max(\min(X_1, X_2), \min(X_3, X_4))$	$(0, \frac{2}{3}, \frac{1}{3}, 0)$
17	$\max(\min(X_1, X_2), \min(X_1, X_3),$ $\min(X_2, X_3, X_4))$	$(0, \frac{2}{3}, \frac{1}{3}, 0)$
18	$\max(\min(X_1, X_2), \min(X_2, X_3),$ $\min(X_3, X_4))$	$(0, \frac{1}{2}, \frac{1}{2}, 0)$
19	$\max(\min(X_1, \max(X_2, X_3, X_4)),$ $\min(X_2, X_3, X_4))$	$(0, \frac{1}{2}, \frac{1}{2}, 0)$
20	$\min(\max(X_1, X_2), \max(X_1, X_3),$ $\max(X_2, X_3, X_4))$	$(0, \frac{1}{3}, \frac{2}{3}, 0)$
21	$\min(\max(X_1, X_2), \max(X_3, X_4))$	$(0, \frac{1}{3}, \frac{2}{3}, 0)$
22	$\min(\max(X_1, X_2), \max(X_1, X_3, X_4),$ $\max(X_2, X_3, X_4))$	$(0, \frac{1}{6}, \frac{5}{6}, 0)$
23	$X_{3:4}$ (2-out-of-4)	$(0, 0, 1, 0)$
24	$\max(X_1, \min(X_2, X_3, X_4))$	$(0, \frac{1}{2}, \frac{1}{4}, \frac{1}{4})$
25	$\max(X_1, \min(X_2, X_3), \min(X_3, X_4))$	$(0, \frac{1}{6}, \frac{7}{12}, \frac{1}{4})$
26	$\max(X_{2:3}, X_4)$	$(0, 0, \frac{3}{4}, \frac{1}{4})$
27	$\min(\max(X_1, X_2, X_3), \max(X_2, X_3, X_4))$	$(0, 0, \frac{1}{2}, \frac{1}{2})$
28	$X_{4:4} = \max(X_1, X_2, X_3, X_4)$ (4-parallel)	$(0, 0, 0, 1)$

when the component lifetimes are IID with a common continuous distribution F. Also note that if T is a system with signature $\mathbf{s} = (s_1, \ldots, s_n)$, then the signature of its dual system T^D is $\mathbf{s}^D = (s_n, \ldots, s_1)$ (see e.g. the systems with numbers 10 and 27). This is a general property and so we do not need to compute the signatures of the dual systems 20-28. The signatures for all the coherent systems with $n = 5$ and $n = 6$ components were obtained in Navarro and Rubio (2010).

As mentioned above, expression (2.5) is a mixture representation for T. So we can use here all the properties for mixtures. For example, the expected lifetime for the system (MTTF) is

$$E(T) = \sum_{i=1}^{n} s_i E(X_{i:n}).$$

A similar property holds for the respective distribution functions and, in the absolutely continuous case, the respective probability density functions (PDF) satisfy

$$f_T(t) = \sum_{i=1}^{n} s_i f_{i:n}(t). \tag{2.11}$$

The PDF of the order statistics can be obtained as follows.

Proposition 2.3 *If $X_1, \ldots, X_n$ are IID$\sim F$ and F is absolutely continuous with PDF f, then the PDF of $X_{i:n}$ is*

$$f_{i:n}(t) = i\binom{n}{i} f(t) F^{i-1}(t) \bar{F}^{n-i}(t). \tag{2.12}$$

Proof From (2.8) we have

$$\bar{F}_{i:n}(t) = \sum_{j=0}^{i-1} \binom{n}{j} F^j(t) \bar{F}^{n-j}(t).$$

Differentiating this expression we obtain

$$\bar{F}'_{i:n}(t) = f(t) \sum_{j=1}^{i-1} \binom{n}{j} j F^{j-1}(t) \bar{F}^{n-j}(t) - f(t) \sum_{j=0}^{i-1} \binom{n}{j} (n-j) F^j(t) \bar{F}^{n-j-1}(t)$$

$$= nf(t) \sum_{j=1}^{i-1} \binom{n-1}{j-1} F^{j-1}(t) \bar{F}^{n-j}(t) - nf(t) \sum_{j=0}^{i-1} \binom{n-1}{j} F^j(t) \bar{F}^{n-j-1}(t)$$

$$= nf(t) \sum_{k=0}^{i-2} \binom{n-1}{k} F^k(t) \bar{F}^{n-k-1}(t) - nf(t) \sum_{j=0}^{i-1} \binom{n-1}{j} F^j(t) \bar{F}^{n-j-1}(t)$$

$$= -nf(t) \binom{n-1}{i-1} F^{i-1}(t) \bar{F}^{n-i}(t)$$

$$= -i\binom{n}{i} f(t) F^{i-1}(t) \bar{F}^{n-i}(t)$$

and so (2.12) holds. □

Remark 2.1 The PDF in (2.12) can be rewritten as

$$f_{i:n}(t) = \frac{\Gamma(n+1)}{\Gamma(i)\Gamma(n-i+1)} f(t) F^{i-1}(t) \bar{F}^{n-i}(t) \tag{2.13}$$

where $\Gamma(p) = \int_0^\infty x^{p-1} e^{-x} dx$ is the gamma function. It can be proved that the function defined in (2.13) for $i, n \in \mathbb{R}$ satisfying $1 \leq i \leq n$ is a proper PDF. Then we can consider the random variable $X_{i:n}$ having this PDF for $i, n \in \mathbb{R}$ satisfying $1 \leq i \leq n$ as an extension of the order statistics (k-out-of-n systems) that are obtained when i and n are integers.

As an immediate consequence of (2.12), the PDF of the system can be obtained as follows.

Corollary 2.1 *If T is the lifetime of a coherent system with IID component lifetimes $X_1, \ldots, X_n$ having a common absolutely continuous distribution function F with PDF f, then the PDF f_T of T can be written as*

$$f_T(t) = \sum_{i=1}^n s_i f_{i:n}(t) = f(t) \sum_{i=1}^n i s_i \binom{n}{i} F^{i-1}(t) \bar{F}^{n-i}(t)$$

for all t.

All these expressions are convex combinations and so, the values for the systems will be between the minimum and maximum values for the k-out-of-n systems. In particular, as $X_{1:n} \leq \cdots \leq X_{n:n}$, then their respective reliability functions are ordered, that is,

$$\bar{F}_{1:n}(t) \leq \cdots \leq \bar{F}_{n:n}(t)$$

for all t. Even more, as $X_{1:n} \leq T \leq X_{n:n}$, then

$$\bar{F}_{1:n}(t) \leq \bar{F}_T(t) \leq \bar{F}_{n:n}(t)$$

for all t. In the IID$\sim F$ case we can be more precise and write

$$\bar{F}_{i:n}(t) \leq \bar{F}_T(t) \leq \bar{F}_{j:n}(t),$$

where i is the smallest index with $s_i > 0$ and j is the greatest index with $s_j > 0$. In the preceding example the signature is $(1/3, 2/3, 0)$. Hence $i = 1$ and $j = 2$ and so the orderings for the reliability functions in Fig. 2.3, left, is a general property for any continuous distribution F. In the next chapter we will use (2.5) and the ordering properties for mixtures given in Shaked and Shanthikumar (2007) to compare two systems by comparing their signatures.

However, the expressions for the mean residual life and the hazard rate functions are different. So it is difficult to determine the behavior of these functions in mixtures. For example, the hazard rate of the system can be written from (2.5) and (2.11) as

$$h_T(t) = \frac{f_T(t)}{\bar{F}_T(t)} = \frac{\sum_{i=1}^n s_i f_{i:n}(t)}{\sum_{i=1}^n s_i \bar{F}_{i:n}(t)} = \sum_{i=1}^n w_i(t) h_{i:n}(t) \tag{2.14}$$

where $h_{i:n} = f_{i:n}/\bar{F}_{i:n}$ is the hazard rate of $X_{i:n}$ and

$$w_i(t) = \frac{s_i \bar{F}_{i:n}(t)}{\sum_{j=1}^{n} s_j \bar{F}_{j:n}(t)}.$$

Note that $0 \le w_i(t) \le 1$ and $\sum_{i=1}^{n} w_i(t) = 1$ for all t. Hence (2.14) is also a convex combination but, in this case, the coefficients $w_1(t), \ldots, w_n(t)$ depend on t.

In the IID case, the hazard rate functions of the k-out-of-n systems are ordered, that is, $h_{1:n} \ge \cdots \ge h_{n:n}$. Hence, in this case, we also have

$$h_{i:n}(t) \ge h_T(t) \ge h_{j:n}(t)$$

for all F and the indices defined above. For example, the hazard rate functions for the system in Example 2.2 when the component lifetimes are IID with a standard exponential are plotted in Fig. 2.3, right. The code in R to get this plot (by using also the code written above) is:

```
f<-function(t) exp(-t)
f1<-function(t) 3*f(t)*(R(t))^2
f2<-function(t) (6*R(t)-6*(R(t))^2)*f(t)
f3<-function(t) (3-6*R(t)+3*(R(t))^2)*f(t)
fT<-function(t) s1*f1(t)+s2*f2(t)+s3*f3(t)
curve(fT(x)/RT(x),0,5,ylim=c(0,3),ylab='HR',col='red',lwd=2)
curve(f1(x)/R1(x),add=T,lwd=2)
curve(f2(x)/R2(x),add=T,lwd=2)
curve(f3(x)/R3(x),add=T,lwd=2)
abline(h=1,lty=2,lwd=2)
```

The following example shows that the continuity assumption in Samaniego's representation cannot be dropped out.

Example 2.3 Let us consider the series system with lifetime $T = X_{1:2} = \min(X_1, X_2)$, where X_1, X_2 are IID with a common Bernoulli distribution of parameter $1/2$, that is, $\Pr(X_i = 1) = \Pr(X_i = 0) = 1/2$ for $i = 1, 2$. Then

$$\Pr(T = X_{1:2}) = 1$$

and

$$\Pr(T = X_{2:2}) = \Pr(X_1 = X_2) = \frac{1}{2}.$$

So (2.5) does not hold with these coefficients. Also note that $1 + 1/2 > 1$. However, the signature computed from the structure by using (2.6) (or (2.7)) is $\mathbf{s} = (1, 0)$. Note that (2.5) holds with these coefficients.   ◄

Therefore, in the general case, that is, when $(X_1, \ldots, X_n)$ is an arbitrary random vector with joint distribution function

$$\mathbf{F}(x_1, \ldots, x_n) = \Pr(X_1 \le x_1, \ldots, X_n \le x_n)$$

and joint reliability function

$$\bar{\mathbf{F}}(x_1, \ldots, x_n) = \Pr(X_1 > x_1, \ldots, X_n > x_n),$$

we can define two signatures as follows.

Definition 2.1 The **structural signature** of a coherent system ψ is $\mathbf{s} = (s_1, \ldots, s_n)$ where s_i is given by (2.6) (or by (2.7)).

Definition 2.2 The **probabilistic signature** of a coherent system with lifetime T and component lifetimes $(X_1, \ldots, X_n)$ is $\mathbf{p} = (p_1, \ldots, p_n)$ where $p_i = \Pr(T = X_{i:n})$.

Clearly, $\mathbf{s}$ only depends on the structure of the system ψ while $\mathbf{p}$ can also depend on the joint distribution function of the components $\mathbf{F}$. In the preceding example $\mathbf{p} = (1, 1/2)$ and $\mathbf{s} = (1, 0)$. In this case Samaniego's representation (2.5) holds for $\mathbf{s}$. We will see that this is true for the general IID case but that we will not have representations for non-ID components.

As Samaniego's representation does not necessarily hold in the general case we need another way to compute the system reliability. It is provided in the following theorem and it is called the minimal path set representation.

Theorem 2.2 (Minimal path set representation) *If T is the lifetime of a coherent (or semi-coherent) system with minimal path sets $P_1, \ldots, P_r$ and component lifetimes $(X_1, \ldots, X_n)$, then*

$$\bar{F}_T(t) = \sum_{i=1}^{r} \bar{F}_{P_i}(t) - \sum_{i=1}^{r-1} \sum_{j=i+1}^{r} \bar{F}_{P_i \cup P_j}(t) + \cdots + (-1)^{r+1} \bar{F}_{P_1 \cup \ldots \cup P_r}(t) \quad (2.15)$$

for all t, where $\bar{F}_P(t) = \Pr(X_P > t)$ and $X_P = \min_{j \in P} X_j$ for $P \subseteq [n]$.

Proof First note that from (2.1), the system lifetime can be written as $T = \max_{1 \leq i \leq r} X_{P_i}$. Then

$$\bar{F}_T(t) = \Pr(T > t) = \Pr\left(\max_{1 \leq i \leq r} X_{P_i} > t\right) = \Pr\left(\cup_{i=1}^{r} \{X_{P_i} > t\}\right).$$

Hence, by using the inclusion-exclusion formula for the union of events, we obtain (2.15) taking into account that

$$\Pr\left(\{X_{P_i} > t\} \cap \{X_{P_j} > t\}\right) = \Pr\left(X_{P_i \cup P_j} > t\right). \quad \blacktriangleleft$$

Note that (2.15) proves that the reliability function of the system is a linear combination of the reliability functions of the series systems obtained from unions of its minimal path sets. However, some coefficients can be negative and so it is not a mixture representation (as (2.5) was). Note that these coefficients sum up to one (take $t \to -\infty$). These representations are called **generalized mixtures** and they contain the usual mixtures (all the coefficients are non-negative) and the negative mixtures (some coefficients are negative). They have some common properties with the usual mixtures. For example, similar expressions hold for the respective distribution and probability density functions.

For the system with lifetime $T = \min(X_1, \max(X_2, X_3))$ and minimal path sets $P_1 = \{1, 2\}$ and $P_2 = \{1, 3\}$, we have

$$
\begin{aligned}
\bar{F}_T(t) &= \Pr\left(\min(X_1, \max(X_2, X_3)) > t\right) \\
&= \Pr\left(\{\min(X_1, X_2) > t\} \cup \{\min(X_1, X_3) > t\}\right) \\
&= \Pr\left(\min(X_1, X_2) > t\right) + \Pr\left(\min(X_1, X_3) > t\right) - \Pr\left(\min(X_1, X_2, X_3) > t\right) \\
&= \bar{F}_{\{1,2\}}(t) + \bar{F}_{\{1,3\}}(t) - \bar{F}_{\{1,2,3\}}(t).
\end{aligned}
\tag{2.16}
$$

The reliability functions of series systems can be computed from the joint reliability function $\bar{\mathbf{F}}$ of the components. For example, in this system, we have

$$
\bar{F}_{\{1,2\}}(t) = \Pr(\min(X_1, X_2) > t) = \Pr(X_1 > t, X_2 > t) = \bar{\mathbf{F}}(t, t, -\infty).
$$

Analogously, $\bar{F}_{\{1,3\}}(t) = \bar{\mathbf{F}}(t, -\infty, t)$ and $\bar{F}_{\{1,2,3\}}(t) = \bar{\mathbf{F}}(t, t, t)$. Therefore,

$$
\bar{F}_T(t) = \bar{\mathbf{F}}(t, t, -\infty) + \bar{\mathbf{F}}(t, -\infty, t) - \bar{\mathbf{F}}(t, t, t).
$$

In the general case, for the series system $X_{\{1,\dots,k\}}$ we have

$$
\bar{F}_{\{1,\dots,k\}}(t) = \Pr(X_1 > t, \dots, X_k > t) = \bar{\mathbf{F}}(t, \dots, t, -\infty, \dots, -\infty)
$$

where t is repeated k times, for $k = 1, \dots, n$. Similarly, for an arbitrary series system X_P with $P \subseteq [n]$ we have

$$
\bar{F}_P(t) = \Pr\left(\min_{j \in P} X_j > t\right) = \bar{\mathbf{F}}(t_1^P, \dots, t_n^P),
$$

where $t_i^P := t$ if $i \in P$ and $t_i^P := -\infty$ if $i \notin P$.

If the component lifetimes are stochastically independent (IND), that is,

$$
\bar{\mathbf{F}}(x_1, \dots, x_n) = \Pr(X_1 > x_1) \dots \Pr(X_n > x_n),
$$

then these expressions can be reduced to

$$
\bar{F}_P(t) = \prod_{j \in P} \Pr(X_j > t) = \prod_{j \in P} \bar{F}_j(t)
$$

and if they are IID then $\bar{F}_P(t) = \bar{F}^{|P|}(t)$ where $|P|$ is the cardinality of P.

For the above system, we get

$$
\bar{F}_T(t) = \bar{F}_1(t) \bar{F}_2(t) + \bar{F}_1(t) \bar{F}_3(t) - \bar{F}_1(t) \bar{F}_2(t) \bar{F}_3(t)
$$

in the IND case and

$$
\bar{F}_T(t) = 2\bar{F}^2(t) - \bar{F}^3(t)
$$

in the IID case. Of course, this last expression coincides with the one obtained from Samaniego's representation. It does not coincide when the components are independent but not identically distributed (INID). For example, in Fig. 2.4 we plot the system reliability and hazard rate functions when the components are independent and have exponential distributions with means $1, 1/2, 1/3$ (black) and a common mean $1/2$ (red). In this example, the system with heterogeneous components is more reliable than the one with homogeneous components. The code is the following:

```
# Reliability functions
R<-function(t) exp(-2*t)
R1<-function(t) exp(-t)
R2<-function(t) exp(-2*t)
R3<-function(t) exp(-3*t)
RT2<-function(t) 2*(R(t))^2-(R(t))^3
RT1<-function(t) R1(t)*R2(t)+R1(t)*R3(t)-R1(t)*R2(t)*R3(t)
curve(RT2(x),0,2,xlab='t',ylab='Reliability',col='red',lwd=2)
curve(RT1(x),add=T,lwd=2)
curve(R1(x),add=T,lty=2,lwd=2)
curve(R2(x),add=T,lty=2,lwd=2)
curve(R3(x),add=T,lty=2,lwd=2)


#Hazard rates
f<-function(t) 2*exp(-2*t)
f1<-function(t) exp(-t)
f2<-function(t) 2*exp(-2*t)
f3<-function(t) 3*exp(-3*t)
fT2<-function(t) (4*R(t)-3*(R(t))^2)*f(t)
fT1<-function(t) {
   f1(t)*R2(t)+R1(t)*f2(t)+f1(t)*R3(t)+R1(t)*f3(t)
   -f1(t)*R2(t)*R3(t)-f2(t)*R1(t)*R3(t)-f3(t)*R2(t)*R1(t)
}
curve(fT2(x)/RT2(x),0,4,col='red',ylim=c(0,4),lwd=2,ylab='HR)
curve(fT1(x)/RT1(x), add=T,lwd=2)
abline(h=1,lty=2,lwd=2)
abline(h=2,lty=2,lwd=2)
abline(h=3,lty=2,lwd=2)
```

As a consequence we obtain the following representation for the IND case.

Corollary 2.2 (Minimal path set representation, IND case) *If T is the lifetime of a coherent (or semi-coherent) system with minimal path sets $P_1, \ldots, P_r$ and independent component lifetimes $X_1, \ldots, X_n$, then*

$$\bar{F}_T(t) = \sum_{i=1}^{r} \prod_{k \in P_i} \bar{F}_k(t) - \sum_{i=1}^{r-1} \sum_{j=i+1}^{r} \prod_{k \in P_i \cup P_j} \bar{F}_k(t) + \cdots + (-1)^{r+1} \prod_{k \in P_1 \cup \cdots \cup P_r} \bar{F}_k(t)$$

$$(2.17)$$

for all t, where $\bar{F}_k(t) = \Pr(X_k > t)$ for $k = 1, \ldots, n$.

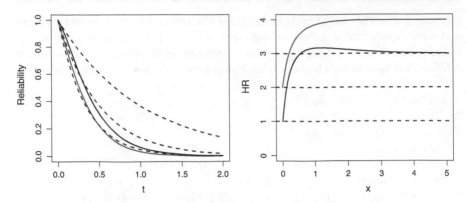

Fig. 2.4 Reliability (left) and hazard rate functions (right) of the system in Example 2.2 when the components are independent and have exponential distributions of means $1, 1/2, 1/3$ (black) and a common mean $1/2$ (red). The dashed lines are the functions for the components

A similar expression can be obtained from the minimal cut sets. It can be stated as follows. Its proof is analogous.

Theorem 2.3 (Minimal cut set representation) *If T is the lifetime of a coherent (or semi-coherent) system with minimal cut sets $C_1, \ldots, C_s$ and component lifetimes $(X_1, \ldots, X_n)$, then*

$$F_T(t) = \sum_{i=1}^{s} F^{C_i}(t) - \sum_{i=1}^{s-1} \sum_{j=i+1}^{s} F^{C_i \cup C_j}(t) + \cdots + (-1)^{s+1} F^{C_1 \cup \ldots \cup C_s}(t) \quad (2.18)$$

for all t, where $F^P(t) = \Pr(X^P \leq t)$ and $X^P = \max_{j \in P} X_j$ for $P \subseteq [n]$.

Note that we obtain again generalized mixtures. Hence the same expression also holds for the respective reliability functions. We use the distribution functions because, in the IND case, it can be reduced to the following corollary.

Corollary 2.3 (Minimal cut set representation, IND case) *If T is the lifetime of a coherent (or semi-coherent) system with minimal cut sets $C_1, \ldots, C_s$ and independent component lifetimes $X_1, \ldots, X_n$, then*

$$F_T(t) = \sum_{i=1}^{s} \prod_{k \in C_i} F_k(t) - \sum_{i=1}^{s-1} \sum_{j=i+1}^{s} \prod_{k \in C_i \cup C_j} F_k(t) + \cdots + (-1)^{r+1} \prod_{k \in C_1 \cup \cdots \cup C_s} F_k(t)$$
$$(2.19)$$

for all t, where $F_k(t) = \Pr(X_k \leq t)$ for $k = 1, \ldots, n$.

The general expressions obtained from the minimal path or cut sets can be reduced to simpler expressions when the component lifetimes are exchangeable. The formal definition is the following.

Definition 2.3 The random vector $(X_1, \ldots, X_n)$ is **exchangeable** (EXC) if

$$(X_1, \ldots, X_n) =_{ST} (X_{\sigma(1)}, \ldots, X_{\sigma(n)})$$

for all the permutations $\sigma : [n] \to [n]$, where $=_{ST}$ denotes equality in distribution (law).

Clearly, $(X_1, \ldots, X_n)$ is EXC iff its joint distribution (or reliability) function $\mathbf{F}$ is permutation symmetric, that is,

$$\mathbf{F}(x_1, \ldots, x_n) =_{ST} \mathbf{F}(x_{\sigma(1)}, \ldots, x_{\sigma(n)})$$

for all the permutations $\sigma : [n] \to [n]$ and all $x_1, \ldots, x_n \in \mathbb{R}$.

If $(X_1, \ldots, X_n)$ is EXC, then all the marginal distributions of dimension k are equal and, in particular, the variables are ID. Hence, the distributions of all the series (or parallel) systems with k components are equal and we have

$$\bar{F}_P(t) = \bar{F}_{\{1,\ldots,k\}}(t) = \bar{\mathbf{F}}(t, \ldots, t, -\infty, \ldots, -\infty),$$

where t is repeated $k = |P|$ times. As a consequence, we obtain the following representation given in Navarro et al. (2007).

Theorem 2.4 (Minimal signature representation) *If T is the lifetime of a coherent (or semi-coherent) system with EXC component lifetimes $(X_1, \ldots, X_n)$, then*

$$\bar{F}_T(t) = \sum_{i=1}^{n} a_i \bar{F}_{1:i}(t) \tag{2.20}$$

for all t, where $a_1, \ldots, a_n$ are some integer coefficients such that $a_1 + \cdots + a_n = 1$ and $\bar{F}_{1:i}(t) = \Pr(\min(X_1, \ldots, X_i) > t)$ for $i = 1, \ldots, n$.

Proof From (2.15), we have that $\bar{F}_T$ is a linear combination of reliability functions $\bar{F}_P$ of series systems. But, if $(X_1, \ldots, X_n)$ is EXC, then $\bar{F}_P$ can be replaced by $\bar{F}_{1:i}$ with $i = |P|$. Hence (2.20) holds for some coefficients $a_1, \ldots, a_n \in \mathbb{Z}$. Moreover these coefficients sum up to one (take $t \to -\infty$). $\qquad\square$

The vector $\mathbf{a} = (a_1, \ldots, a_n)$ with these coefficients was call the **minimal signature** of the system in Navarro et al. (2007). A similar representation can be obtained by using the parallel systems as follows.

Theorem 2.5 (Maximal signature representation) *If T is the lifetime of a coherent (or semi-coherent) system with EXC component lifetimes $(X_1, \ldots, X_n)$, then*

$$F_T(t) = \sum_{i=1}^{n} b_i F_{i:i}(t) \tag{2.21}$$

for all t, where $b_1, \ldots, b_n$ are some integer coefficients such that $b_1 + \cdots + b_n = 1$ and $F_{i:i}(t) = \Pr(\max(X_1, \ldots, X_i) \leq t)$ for $i = 1, \ldots, n$.

The vector $\mathbf{b} = (b_1, \ldots, b_n)$ with these coefficients was call the **maximal signa-ture** of the system in Navarro et al. (2007). Note that both representations (2.20) and (2.21) are generalized mixtures and that they hold for the general EXC case (including discrete or singular distributions). In both we can use distribution or reliability functions. However, it is better to use reliability functions with series systems and distribution functions with parallel systems. In the absolutely continuous case, we can also use probability density functions. We will see later that they do not hold without the EXC assumption.

Let us see an example. For the coherent system with lifetime $T = \min(X_1, \max(X_2, X_3))$, we get

$$\bar{F}_T(t) = \bar{F}_{\{1,2\}}(t) + \bar{F}_{\{1,3\}}(t) - \bar{F}_{\{1,2,3\}}(t)$$

that, in the EXC case, can be reduced to

$$\bar{F}_T(t) = 2\bar{F}_{1:2}(t) - \bar{F}_{1:3}(t).$$

Hence its minimal signature is $\mathbf{a} = (0, 2, -1)$. To compute its maximal signature we first write its distribution function as

$$F_T(t) = F^{\{1\}}(t) + F^{\{2,3\}}(t) - F^{\{1,2,3\}}(t)$$

which, in the EXC case, gives

$$F_T(t) = F_{1:1}(t) + F_{2:2}(t) - F_{3:3}(t)$$

for all t. So its maximal signature is $\mathbf{b} = (1, 1, -1)$.

The minimal and maximal signatures of all the coherent systems with 1-4 components are given in Table 2.2. It can be proved that the minimal (maximal) signature of a system coincides with the maximal (minimal) signature of its dual system (see, e.g., the systems in rows 10 and 27). This property is due to the fact that the minimal cut (path) sets of a system are the minimal path (cut) sets of its dual system.

The EXC case includes the general IID case and so we can obtain the following representations.

Theorem 2.6 (Minimal and maximal signature representations, IID case) *If T is the lifetime of a coherent (or semi-coherent) system with minimal and maximal signatures $(a_1, \ldots, a_n)$ and $(b_1, \ldots, b_n)$ and IID$\sim F$ component lifetimes, then*

$$\bar{F}_T(t) = \sum_{i=1}^{n} a_i \bar{F}^i(t) \tag{2.22}$$

and

$$F_T(t) = \sum_{i=1}^{n} b_i F^i(t) \tag{2.23}$$

for all t.

Table 2.2 Minimal **a** and maximal **b** signatures of all the coherent systems with 1-4 exchangeable components

i	T_i	**a**	**b**
1	$X_{1:1} = X_1$	(1)	(1)
2	$X_{1:2} = \min(X_1, X_2)$ (2-series)	$(0, 1)$	$(2, -1)$
3	$X_{2:2} = \max(X_1, X_2)$ (2-parallel)	$(2, -1)$	$(0, 1)$
4	$X_{1:3} = \min(X_1, X_2, X_3)$ (3-series)	$(0, 0, 1)$	$(3, -3, 1)$
5	$\min(X_1, \max(X_2, X_3))$	$(0, 2, -1)$	$(1, 1, -1)$
6	$X_{2:3}$ (2-out-of-3)	$(0, 3, -2)$	$(0, 3, -2)$
7	$\max(X_1, \min(X_2, X_3))$	$(1, 1, -1)$	$(0, 2, -1)$
8	$X_{3:3} = \max(X_1, X_2, X_3)$ (3-parallel)	$(3, -3, 1)$	$(0, 0, 1)$
9	$X_{1:4} = \min(X_1, X_2, X_3, X_4)$ (series)	$(0, 0, 0, 1)$	$(4, -6, 4, -1)$
10	$\max(\min(X_1, X_2, X_3), \min(X_2, X_3, X_4))$	$(0, 0, 2, -1)$	$(2, 0, -2, 1)$
11	$\min(X_{2:3}, X_4)$	$(0, 0, 3, -2)$	$(1, 3, -5, 2)$
12	$\min(X_1, \max(X_2, X_3), \max(X_3, X_4))$	$(0, 1, 1, -1)$	$(1, 2, -3, 1)$
13	$\min(X_1, \max(X_2, X_3, X_4))$	$(0, 3, -3, 1)$	$(1, 0, 1, -1)$
14	$X_{2:4}$ (3-out-of-4)	$(0, 0, 4, -3)$	$(0, 6, -8, 3)$
15	$\max(\min(X_1, X_2), \min(X_1, X_3, X_4),$ $\min(X_2, X_3, X_4))$	$(0, 1, 2, -2)$	$(0, 5, -6, 2)$
16	$\max(\min(X_1, X_2), \min(X_3, X_4))$	$(0, 2, 0, -1)$	$(0, 4, -4, 1)$
17	$\max(\min(X_1, X_2), \min(X_1, X_3),$ $\min(X_2, X_3, X_4))$	$(0, 2, 0, -1)$	$(0, 4, -4, 1)$
18	$\max(\min(X_1, X_2), \min(X_2, X_3),$ $\min(X_3, X_4))$	$(0, 3, -2, 0)$	$(0, 3, -2, 0)$
19	$\max(\min(X_1, \max(X_2, X_3, X_4)),$ $\min(X_2, X_3, X_4))$	$(0, 3, -2, 0)$	$(0, 3, -2, 0)$
20	$\min(\max(X_1, X_2), \max(X_1, X_3),$ $\max(X_2, X_3, X_4))$	$(0, 4, -4, 1)$	$(0, 2, 0, -1)$
21	$\min(\max(X_1, X_2), \max(X_3, X_4))$	$(0, 4, -4, 1)$	$(0, 2, 0, -1)$
22	$\min(\max(X_1, X_2), \max(X_1, X_3, X_4),$ $\max(X_2, X_3, X_4))$	$(0, 5, -6, 2)$	$(0, 1, 2, -2)$
23	$X_{3:4}$ (2-out-of-4)	$(0, 6, -8, 3)$	$(0, 0, 4, -3)$
24	$\max(X_1, \min(X_2, X_3, X_4))$	$(1, 0, 1, -1)$	$(0, 3, -3, 1)$
25	$\max(X_1, \min(X_2, X_3), \min(X_3, X_4))$	$(1, 2, -3, 1)$	$(0, 1, 1, -1)$
26	$\max(X_{2:3}, X_4)$	$(1, 3, -5, 2)$	$(0, 0, 3, -2)$
27	$\min(\max(X_1, X_2, X_3), \max(X_2, X_3, X_4))$	$(2, 0, -2, 1)$	$(0, 0, 2, -1)$
28	$X_{4:4} = \max(X_1, X_2, X_3, X_4)$ (4-parallel)	$(4, -6, 4, -1)$	$(0, 0, 0, 1)$

The proof is immediate from (2.20) and (2.21) since, in the IID case, we have $\bar{F}_{1:i}(t) = \bar{F}^i(t)$ and $F_{i:i}(t) = F^i(t)$ for all t. So it is better to use reliability functions with the minimal signature and distribution functions with the maximal signature. Note that we do not need the assumption "F is continuous". In the absolutely continuous IID case, the system PDF is

$$f_T(t) = f(t) \sum_{i=1}^{n} i a_i \bar{F}^{i-1}(t) = f(t) \sum_{i=1}^{n} i b_i F^{i-1}(t)$$

where $f = F' = -\bar{F}'$ is the common PDF of the components.

The minimal (or maximal) signature representation can be used to extend Samaniego's representation to the general EXC case (which includes the IID case with a general distribution F). It is stated in the following theorem. This result was obtained in Navarro et al. (2008) by using a different proof. A similar result was obtained previously in Navarro and Rychlik (2007) for absolutely continuous EXC distributions.

Theorem 2.7 (Signature representation, EXC case) *If T is the lifetime of a coherent system with structural signature $(s_1, \ldots, s_n)$ and with EXC component lifetimes, then*

$$\bar{F}_T(t) = \sum_{i=1}^{n} s_i \bar{F}_{i:n}(t) \tag{2.24}$$

for all t.

Proof From (2.20) we have

$$\bar{F}_T(t) = \sum_{i=1}^{n} a_i \bar{F}_{1:i}(t)$$

where $\mathbf{a} = (a_1, \ldots, a_n)$ is the minimal signature of T. This representation can be written as

$$\bar{F}_T(t) = \mathbf{a}(\bar{F}_{1:1}(t), \ldots, \bar{F}_{1:n}(t))',$$

where $\mathbf{v}'$ represents the transpose of $\mathbf{v}$.

We can apply this representation to the k-out-of-n systems as well. Thus, for $X_{1:n}$, which only has a minimal path set $P_1 = \{1, \ldots, n\}$ (the unique set with cardinality n), we obtain the trivial representation

$$\bar{F}_{1:n}(t) = 0\bar{F}_{1:1}(t) + \cdots + 0\bar{F}_{1:n-1}(t) + 1\bar{F}_{1:n}(t),$$

that is, its minimal signature is $(0, \ldots, 0, 1)$.

Analogously, for $X_{2:n}$, its minimal path sets are all the sets with cardinality $n - 1$ (its has $\binom{n}{n-1} = n$ minimal path sets). Then its minimal path set representation is

$$\bar{F}_{2:n}(t) = 0\bar{F}_{1:1}(t) + \cdots + 0\bar{F}_{1:n-2}(t) + n\bar{F}_{1:n-1}(t) - (n-1)\bar{F}_{1:n}(t),$$

that is, its minimal signature is $(0, \ldots, 0, n, -n + 1)$ (the last coefficient is $-n + 1$ because their sum is one).

In general, for $X_{i:n}$, we obtain

$$\bar{F}_{i:n}(t) = 0\bar{F}_{1:1}(t) + \cdots + 0\bar{F}_{1:n-i}(t) + \binom{n}{i}\bar{F}_{1:n-i+1}(t) + \cdots + a_{i,n}\bar{F}_{1:n}(t)$$

for $i = 1, \ldots, n$. The coefficients in that representation are well known in the order statistics literature (see David and Nagaraja 2003, p. 46 or (2.25) below). However, we do not need them. We just need the fact that

$$(\bar{F}_{1:n}(t), \ldots, \bar{F}_{n:n}(t))' = A_n(\bar{F}_{1:1}(t), \ldots, \bar{F}_{1:n}(t))'$$

for all t, where $A_n = (a_{i,j})$ is a triangular non-singular matrix of real (integer) numbers. Hence

$$(\bar{F}_{1:1}(t), \ldots, \bar{F}_{1:n}(t))' = A_n^{-1}(\bar{F}_{1:n}(t), \ldots, \bar{F}_{n:n}(t))'$$

for all t, where A_n^{-1} is the inverse matrix of A_n.

Therefore, by using the minimal signature representation obtained in (2.20), we get

$$\begin{aligned}
\bar{F}_T(t) &= \mathbf{a}(\bar{F}_{1:1}(t), \ldots, \bar{F}_{1:n}(t))' \\
&= \mathbf{a}A_n^{-1}(\bar{F}_{1:n}(t), \ldots, \bar{F}_{n:n}(t))' \\
&= \mathbf{c}(\bar{F}_{1:n}(t), \ldots, \bar{F}_{n:n}(t))' \\
&= \sum_{i=1}^{n} c_i \bar{F}_{i:n}(t)
\end{aligned}$$

for all t, where $\mathbf{c} = (c_1, \ldots, c_n) := \mathbf{a}A_n^{-1}$ are some coefficients that do not depend on the joint distribution of the component lifetimes (they only depend on $\mathbf{a}$ and A_n).

In the IID continuous case these coefficients coincide with the structural signature coefficients (take e.g. $F(t) = t$ for $0 \le t \le 1$), that is, $\mathbf{c} = \mathbf{s}$ and so (2.24) holds. $\square$

Remark 2.2 The preceding theorem proves that $\bar{F}_T$ belongs to the vectorial space generated by the reliability functions of the k-out-of-n systems which coincides with the one generated by the series system reliability functions (in the EXC case). Actually, in many cases, these reliability functions are bases of this space and so the signatures can be seen as the coordinates of $\bar{F}_T$ in these bases. Thus the structural signature can be obtained from the minimal signature as

$$\mathbf{s} = \mathbf{a}A_n^{-1}$$

and vice versa

$$\mathbf{a} = \mathbf{s}A_n.$$

Moreover, it can be proved that, in the absolutely continuous EXC case, $s_i = \Pr(T = X_{i:n})$ (i.e. the structural and probability signatures coincide), see Navarro and Rychlik (2007).

Note that the rows of A_n are the minimal signatures of the k-out-of-n systems. As mentioned in the proof, the coefficients in A_n are well known in the order statistics literature. Actually, from David and Nagaraja (2003), p. 46, we have

$$\bar{F}_{i:n}(t) = \sum_{j=n-i+1}^{n} (-1)^{j-n+i-1} \binom{n}{j} \binom{j-1}{n-i} \bar{F}_{1:j}(t). \tag{2.25}$$

Of course, the preceding theorem can also be applied to the case of IID$\sim F$ component lifetimes. Here we do not the continuity assumption for F but we have to use the structural signature (not the probabilistic signature).

Remark 2.3 A similar proof can be obtained by using the maximal signature representations of the k-out-of-n systems. So we can also write

$$\mathbf{s} = \mathbf{b} B_n^{-1}$$

and

$$\mathbf{b} = \mathbf{s} B_n$$

for a triangular non-singular matrix $B_n = (b_{i,j})$. The coefficients in B_n can also be obtained from David and Nagaraja (2003), p. 46, as

$$\bar{F}_{i:n}(t) = \sum_{j=i}^{n} b_{i,j} \bar{F}_{j:j}(t) = \sum_{j=i}^{n} (-1)^{j-i} \binom{n}{j} \binom{j-1}{i-1} \bar{F}_{j:j}(t). \tag{2.26}$$

The rows of B_n are the maximal signatures of the order statistics. Note that, as a consequence, $\mathbf{a}$ can be computed from $\mathbf{b}$ and vice versa through

$$\mathbf{b} = \mathbf{s} B_n = \mathbf{a} A_n^{-1} B_n = \mathbf{a} C_n$$

and

$$\mathbf{a} = \mathbf{s} A_n = \mathbf{b} B_n^{-1} A_n = \mathbf{b} C_n^{-1},$$

where $C_n = A_n^{-1} B_n$. If one prefer to use column vectors, just take the transposed matrices.

Example 2.4 Let us obtain A_3 without using (2.25). As mentioned in the proof, the (trivial) minimal signature representation for $X_{1:3}$ is

$$\bar{F}_{1:3}(t) = 0\bar{F}_{1:1}(t) + 0\bar{F}_{1:2}(t) + 1\bar{F}_{1:3}(t).$$

Analogously, for $X_{2:3}$, we have

$$\bar{F}_{2:3}(t) = 0\bar{F}_{1:1}(t) + 3\bar{F}_{1:2}(t) - 2\bar{F}_{1:3}(t).$$

Finally, the minimal path set representation for $X_{3:3}$ in the EXC case is

$$\bar{F}_{3:3}(t) = 3\bar{F}_{1:1}(t) - 3\bar{F}_{1:2}(t) + 1\bar{F}_{1:3}(t).$$

Hence

$$\begin{pmatrix} \bar{F}_{1:3}(t) \\ \bar{F}_{2:3}(t) \\ \bar{F}_{3:3}(t) \end{pmatrix} = \begin{pmatrix} 0 & 0 & 1 \\ 0 & 3 & -2 \\ 3 & -3 & 1 \end{pmatrix} \begin{pmatrix} \bar{F}_{1:1}(t) \\ \bar{F}_{1:2}(t) \\ \bar{F}_{1:3}(t) \end{pmatrix},$$

that is,

$$A_3 = \begin{pmatrix} 0 & 0 & 1 \\ 0 & 3 & -2 \\ 3 & -3 & 1 \end{pmatrix}.$$

Its inverse matrix is

$$A_3^{-1} = \begin{pmatrix} 1/3 & 1/3 & 1/3 \\ 2/3 & 1/3 & 0 \\ 1 & 0 & 0 \end{pmatrix}.$$

These matrices can be used to obtain one signature from the other. For example, for the system with lifetime $T = \min(X_1, \max(X_2, X_3))$, $\mathbf{s}$ can be computed from $\mathbf{a}$ as

$$(s_1, s_2, s_3) = (a_1, a_2, a_3) A_3^{-1} = (0, 2, -1) \begin{pmatrix} 1/3 & 1/3 & 1/3 \\ 2/3 & 1/3 & 0 \\ 1 & 0 & 0 \end{pmatrix} = (1/3, 2/3, 0).$$

Conversely, $\mathbf{a}$ can be computed from $\mathbf{s}$ through

$$(a_1, a_2, a_3) = (s_1, s_2, s_3) A_3 = (1/3, 2/3, 0) \begin{pmatrix} 0 & 0 & 1 \\ 0 & 3 & -2 \\ 3 & -3 & 1 \end{pmatrix} = (0, 2, -1).$$

Analogously, we obtain

$$B_3 = \begin{pmatrix} 3 & -3 & 1 \\ 0 & 3 & -2 \\ 0 & 0 & 1 \end{pmatrix}.$$

Note that the rows of B_n are the rows of A_n in the reverse order (since the dual system of $X_{i:n}$ is $X_{n-i+1:n}$). Hence the maximal signature of the system can be obtained as

$$(b_1, b_2, b_3) = (s_1, s_2, s_3) B_3 = (1/3, 2/3, 0) \begin{pmatrix} 3 & -3 & 1 \\ 0 & 3 & -2 \\ 0 & 0 & 1 \end{pmatrix} = (1, 1, -1).$$

It can be obtained directly from the minimal signature as

$$(b_1, b_2, b_3) = (a_1, a_2, a_3) A_3^{-1} B_3 = (0, 2, -1) \begin{pmatrix} 1 & 0 & 0 \\ 2 & -1 & 0 \\ 3 & -3 & 1 \end{pmatrix} = (1, 1, -1).$$

Note that the rows of $C_3 = A_3^{-1} B_3$ are the maximal signatures of the series systems $X_{1:1}, X_{1:2}, X_{1:3}$. Analogously,

$$(a_1, a_2, a_3) = (b_1, b_2, b_3) B_3^{-1} A_3 = (1, 1, -1) \begin{pmatrix} 1 & 0 & 0 \\ 2 & -1 & 0 \\ 3 & -3 & 1 \end{pmatrix} = (0, 2, -1)$$

that is, $C_3^{-1} = C_3$. This is a general property, that is, $C_n^{-1} = C_n$ for all n.   ◄

The Samaniego's representation can also be extended to semi-coherent systems as follows. This results was obtained in Navarro et al. (2008) (by using a different proof).

Theorem 2.8 *If T is the lifetime of a coherent system with component lifetimes $X_1, \ldots, X_k$ contained in an EXC random vector $(X_1, \ldots, X_n)$ $(k < n)$, then*

$$\bar{F}_T(t) = \sum_{i=1}^{n} s_i^{(n)} \bar{F}_{i:n}(t) \tag{2.27}$$

for all t, where $s_1^{(n)}, \ldots, s_n^{(n)}$ are some coefficients that only depend on the structure of the system and that satisfy $s_1^{(n)} + \cdots + s_n^{(n)} = 1$.

Proof As $(X_1, \ldots, X_n)$ is EXC, so is $(X_1, \ldots, X_k)$. Hence, from (2.20), we have

$$\bar{F}_T(t) = \sum_{i=1}^{k} a_i \bar{F}_{1:k}(t),$$

where $\mathbf{a} = (a_1, \ldots, a_k)$ is the minimal signature of T. This representation can be written as

$$\bar{F}_T(t) = \mathbf{a}^{(n)} (\bar{F}_{1:1}(t), \ldots, \bar{F}_{1:n}(t))'$$

where $\mathbf{a}^{(n)} := (a_1, \ldots, a_k, 0, \ldots, 0) \in \mathbb{Z}^n$.

We can also apply here the representation for the k-out-of-n systems obtained in the preceding theorem. Thus,

$$(\bar{F}_{1:n}(t), \ldots, \bar{F}_{n:n}(t))' = A_n (\bar{F}_{1:1}(t), \ldots, \bar{F}_{1:n}(t))'$$

for all t, where A_n is a triangular non-singular matrix of real (integer) numbers. Hence

$$(\bar{F}_{1:1}(t), \ldots, \bar{F}_{1:n}(t))' = A_n^{-1} (\bar{F}_{1:n}(t), \ldots, \bar{F}_{n:n}(t))'$$

for all t, where A_n^{-1} is the inverse matrix of A_n.

Therefore, by using the representation obtained above, we get

$$\begin{aligned}
\bar{F}_T(t) &= \mathbf{a}^{(n)} (\bar{F}_{1:1}(t), \ldots, \bar{F}_{1:n}(t))' \\
&= \mathbf{a}^{(n)} A_n^{-1} (\bar{F}_{1:n}(t), \ldots, \bar{F}_{n:n}(t))' \\
&= \mathbf{s}^{(n)} (\bar{F}_{1:n}(t), \ldots, \bar{F}_{n:n}(t))' \\
&= \sum_{i=1}^{n} s_i^{(n)} \bar{F}_{i:n}(t)
\end{aligned}$$

for all t, where $\mathbf{s}^{(n)} = (s_1^{(n)}, \ldots, s_n^{(n)}) := \mathbf{a}^{(n)} A_n^{-1}$ are some coefficients that do not depend on the joint distribution of the component lifetimes (they only depend on $\mathbf{a}$ and A_n) and that satisfy $s_1^{(n)} + \cdots + s_n^{(n)} = 1$. $\square$

The vector $\mathbf{s}^{(n)} = (s_1^{(n)}, \ldots, s_n^{(n)})$ is called the **structural signature of order n** of the system. It can be proved that if $(X_1, \ldots, X_n)$ has an absolutely continuous EXC distribution, then $s_i^{(n)} = \Pr(T = X_{i:n})$. Hence $s_i^{(n)} \geq 0$ and so (2.27) is a mixture representation. The structural signature of order n of a semi-coherent system ψ can also be computed from

$$s_i^{(n)} = \frac{1}{\binom{n}{i-1}} \sum_{\sum_{j=1}^{n} x_j = n-i+1} \psi(x_1, \ldots x_n) - \frac{1}{\binom{n}{i}} \sum_{\sum_{j=1}^{n} x_j = n-i} \psi(x_1, \ldots x_n)$$

for $i = 1, \ldots, n$. Of course, if ψ is a coherent system of order n, then we obtain the expression of the structural signature given in (2.7).

Analogously, $\mathbf{a}^{(n)}. = (a_1, \ldots, a_k, 0, \ldots, 0)$ can be called the **minimal signature of order n**. Note that $\mathbf{s}^{(n)} = \mathbf{a}^{(n)} A_n^{-1}$ and $\mathbf{a}^{(n)} = \mathbf{s}^{(n)} A_n$. The **maximal signature of order n** can be defined in a similar way as $\mathbf{b}^{(n)} = (b_1, \ldots, b_k, 0, \ldots, 0)$. It can be used to obtain an alternative proof for the preceding theorem with $\mathbf{s}^{(n)} = \mathbf{b}^{(n)} B_n^{-1}$ and $\mathbf{b}^{(n)} = \mathbf{s}^{(n)} B_n$.

Remark 2.4 The preceding theorem can also be obtained by using the "**Triangle Rule**" of the order statistics. Thus, if $(X_1, \ldots, X_{n+1})$ are EXC without ties, then

$$\Pr(X_{i:n} < X_{n+1} < X_{i+1:n}) = \Pr(X_{n+1} = X_{i+1:n}) = \frac{1}{n+1}$$

for $i = 0, \ldots, n$ where, by convention $X_{0:n} = -\infty$ and $X_{n+1:n} = \infty$. Hence

$$\Pr(X_{i:n} = X_{i+1:n+1}) = \Pr(X_{n+1} < X_{i:n}) = \frac{i}{n+1}$$

and so

$$\Pr(X_{i:n} = X_{i:n+1}) = 1 - \frac{i}{n+1} = \frac{n+1-i}{n+1}.$$

Consequently the order statistics from an EXC random vector without ties satisfy the following triangle rule

$$\bar{F}_{i:n}(t) = \frac{n+1-i}{n+1} \bar{F}_{i:n+1}(t) + \frac{i}{n+1} \bar{F}_{i+1:n+1}(t) \tag{2.28}$$

for all t. Note that we can use this expression in (2.24) to write the reliability function $\bar{F}_T$ of a coherent system with n components as a linear combination of $\bar{F}_{1:n+1}, \ldots, \bar{F}_{n+1:n+1}$, that is, to compute its signature of order $n+1$. Thus, if T has the signature $(s_1^{(n)}, \ldots, s_n^{(n)})$ of order n, then

$$\bar{F}_T = \sum_{i=1}^{n} s_i^{(n)} \bar{F}_{i:n}$$

$$= \sum_{i=1}^{n} s_i^{(n)} \frac{n+1-i}{n+1} \bar{F}_{i:n+1} + \sum_{i=1}^{n} s_i^{(n)} \frac{i}{n+1} \bar{F}_{i+1:n+1}$$

$$= \sum_{i=1}^{n} s_i^{(n)} \frac{n+1-i}{n+1} \bar{F}_{i:n+1} + \sum_{i=2}^{n+1} s_{i-1}^{(n)} \frac{i-1}{n+1} \bar{F}_{i:n+1}$$

$$= \frac{n s_1^{(n)}}{n+1} \bar{F}_{1:n+1} + \sum_{i=2}^{n} \left(\frac{i-1}{n+1} s_{i-1}^{(n)} + \frac{n+1-i}{n+1} s_i^{(n)} \right) \bar{F}_{i:n+1} + \frac{n s_n^{(n)}}{n+1} \bar{F}_{n+1:n+1}.$$

Table 2.3 Signatures of order 4 of all the coherent systems with 1-3 EXC components

	T	$\mathbf{s}^{(4)}$
1	$X_{1:1} = X_1$	$(\frac{1}{4}, \frac{1}{4}, \frac{1}{4}, \frac{1}{4})$
2	$X_{1:2} = \min(X_1, X_2)$ (2-series)	$(\frac{1}{2}, \frac{1}{3}, \frac{1}{6}, 0)$
3	$X_{2:2} = \max(X_1, X_2)$ (2-parallel)	$(0, \frac{1}{6}, \frac{1}{3}, \frac{1}{2})$
4	$X_{1:3} = \min(X_1, X_2, X_3)$ (3-series)	$(\frac{3}{4}, \frac{1}{4}, 0, 0)$
5	$\min(X_1, \max(X_2, X_3))$	$(\frac{1}{4}, \frac{5}{12}, \frac{1}{3}, 0)$
6	$X_{2:3}$ (2-out-of-3)	$(0, \frac{1}{2}, \frac{1}{2}, 0)$
7	$\max(X_1, \min(X_2, X_3))$	$(0, \frac{1}{3}, \frac{5}{12}, \frac{1}{4})$
8	$X_{3:3} = \max(X_1, X_2, X_3)$ (3-parallel)	$(0, 0, \frac{1}{4}, \frac{3}{4})$

Hence, the signature of order $n + 1$ can be obtained as

$$\mathbf{s}^{(n+1)} = \left(\frac{n}{n+1} s_1^{(n)}, \frac{1}{n+1} s_1^{(n)} + \frac{n-1}{n+1} s_2^{(n)}, \frac{2}{n+1} s_2^{(n)} + \frac{n-2}{n+1} s_3^{(n)}, \ldots, \frac{n}{n+1} s_n^{(n)} \right),$$
(2.29)

that is,

$$s_i^{(n+1)} = \frac{i-1}{n+1} s_{i-1}^{(n)} + \frac{n+1-i}{n+1} s_i^{(n)}$$

for $i = 1, \ldots, n + 1$ where, by convention, $s_0^{(n)} = s_{n+1}^{(n)} = 0$. This gives us an alternative proof of Theorem 2.8 based on the Triangle Rule. Actually, this was the proof used in Navarro et al. (2008). We can go further and compute the signature of order n from the signature of order $k < n$. The explicit expressions can be seen in Navarro et al. (2008). Alternatively, we can use (2.29) $n - k$ times.

The signatures of order 4 for all the coherent systems with 1-3 EXC components are given in Table 2.3. Let us see in some examples how to compute them.

Example 2.5 We have seen that if (X_1, X_2) are EXC (or just ID), then

$$\bar{F}_{2:2} = 2\bar{F}_{1:1} - \bar{F}_{1:2}.$$

Hence

$$\bar{F}_{1:1} = \frac{1}{2} \bar{F}_{1:2} + \frac{1}{2} \bar{F}_{2:2},$$

that is, the signature of order 2 of X_1 is $(1/2, 1/2)$. It can also be obtained from the Triangle Rule as follows. Obviously, the signature (of order 1) of X_1 is $\mathbf{s} = (1)$. Hence, from (2.29), we have

$$\mathbf{s}^{(2)} = \left(\frac{1}{2} 1, \frac{1}{2} 1 \right) = \left(\frac{1}{2}, \frac{1}{2} \right).$$

By applying (2.29) again, we get

$$\mathbf{s}^{(3)} = \left(\frac{2}{3} \frac{1}{2}, \frac{1}{3} \frac{1}{2} + \frac{1}{3} \frac{1}{2}, \frac{2}{3} \frac{1}{2} \right) = \left(\frac{1}{3}, \frac{1}{3}, \frac{1}{3} \right).$$

Analogously, if $(X_1, \ldots, X_n)$ are EXC without ties, then $\Pr(X_1 = X_{i:n}) = 1/n$ for $i = 1, \ldots, n$. Hence, the signature of order n of X_1 (or X_i) is $\mathbf{s}^{(n)} = (1/n, \ldots, 1/n)$. ◀

Example 2.6 Let us consider again the coherent system $T = \min(X_1, \max(X_2, X_3))$ with three EXC components. Recall that from Tables 2.1 and 2.2, the signature and the minimal signature of this system are $(1/3, 2/3, 0)$ and $(0, 2, -1)$, respectively. Therefore, the signature of order 4 can be obtained as

$$
\mathbf{s}^{(4)} = \mathbf{a}^{(4)} A_4^{-1} = (0, 2, -1, 0)
\begin{pmatrix}
\frac{1}{4} & \frac{1}{4} & \frac{1}{4} & \frac{1}{4} \\
\frac{1}{2} & \frac{1}{3} & \frac{1}{6} & 0 \\
\frac{3}{4} & \frac{1}{4} & 0 & 0 \\
1 & 0 & 0 & 0
\end{pmatrix}
= \left(\frac{1}{4}, \frac{5}{12}, \frac{1}{3}, 0 \right),
$$

where $(0, 2, -1, 0)$ is the minimal signature of order 4, that is, the coefficients needed to write $\bar{F}_T$ in terms of $\bar{F}_{1:i}$, $i = 1, 2, 3, 4$, and the matrix is the inverse matrix of

$$
A_4 =
\begin{pmatrix}
0 & 0 & 0 & 1 \\
0 & 0 & 4 & -3 \\
0 & 6 & -8 & 3 \\
4 & -6 & 4 & -1
\end{pmatrix}
$$

obtained by placing in the rows the minimal signatures of the order statistics $X_{1:4}, X_{2:4}, X_{3:4}, X_{4:4}$. Note that the rows of A_4^{-1} contain the signatures of order 4 of the series systems $X_{1:1}, X_{1:2}, X_{1:3}, X_{1:4}$.

Another option is to use the following representation based on the signature $(1/3, 2/3, 0)$,

$$
\bar{F}_T(t) = \frac{1}{3}\bar{F}_{1:3}(t) + \frac{2}{3}\bar{F}_{2:3}(t) \tag{2.30}
$$

and the relations of the distributions of order statistics based on the Triangle Rule given in (2.28). Using this rule we have

$$
\bar{F}_{1:3}(t) = \frac{3}{4}\bar{F}_{1:4}(t) + \frac{1}{4}\bar{F}_{2:4}(t)
$$

$$
\bar{F}_{2:3}(t) = \frac{1}{2}\bar{F}_{2:4}(t) + \frac{1}{2}\bar{F}_{3:4}(t)
$$

and replacing these expressions in (2.30), we obtain the signature of order 4 as follows

$$
\bar{F}_T(t) = \frac{1}{3}\bar{F}_{1:3}(t) + \frac{2}{3}\bar{F}_{2:3}(t)
$$

$$
= \frac{1}{3}\left(\frac{3}{4}\bar{F}_{1:4}(t) + \frac{1}{4}\bar{F}_{2:4}(t) \right) + \frac{2}{3}\left(\frac{1}{2}\bar{F}_{2:4}(t) + \frac{1}{2}\bar{F}_{3:4}(t) \right)
$$

$$
= \frac{1}{4}\bar{F}_{1:4}(t) + \frac{5}{12}\bar{F}_{2:4}(t) + \frac{1}{3}\bar{F}_{3:4}(t).
$$

Another option is to apply formula (2.29) to $(1/3, 2/3, 0)$ to get

$$
\left(\frac{3}{4}\frac{1}{3}, \frac{1}{4}\frac{1}{3} + \frac{3}{4}\frac{2}{3}, \frac{2}{4}\frac{2}{3} + \frac{2}{4}0, \frac{3}{4}0 \right) = \left(\frac{1}{4}, \frac{5}{12}, \frac{1}{3}, 0 \right).
$$

◀

We conclude this section with an example extracted from Example 5.1 in Navarro et al. (2008) which proves that representation (2.24) does not necessarily hold without the EXC (ID) assumption. Actually, it proves that the distribution of a system is not necessarily a mixture of the distributions of the order statistics associated to its component lifetimes.

Example 2.7 Let us consider the coherent system with three IND components and with lifetime $T = \min(X_1, \max(X_2, X_3))$. Recall that the minimal path sets of T are $\{1, 2\}$ and $\{1, 3\}$, and so the reliability function of this system can be written as

$$\bar{F}_T(t) = \bar{F}_{\{1,2\}}(t) + \bar{F}_{\{1,3\}}(t) - \bar{F}_{1:3}(t).$$

If we assume that the component lifetimes are IND then

$$\bar{F}_T(t) = \bar{F}_1(t)\bar{F}_2(t) + \bar{F}_1(t)\bar{F}_3(t) - \bar{F}_1(t)\bar{F}_2(t)\bar{F}_3(t).$$

However, in general, we do not know if $\bar{F}_T$ can necessarily be written as a mixture of $\bar{F}_{1:3}$, $\bar{F}_{2:3}$ and $\bar{F}_{3:3}$. For example, if the components have exponential distributions with means $1/2$, 1 and 1, respectively, then

$$\bar{F}_1(t) = e^{-2t}$$
$$\bar{F}_2(t) = \bar{F}_3(t) = e^{-t}$$
$$\bar{F}_{\{1,2\}}(t) = \bar{F}_{\{1,3\}}(t) = e^{-3t},$$
$$\bar{F}_{1:3}(t) = e^{-4t},$$
$$\bar{F}_{2:3}(t) = e^{-2t} + 2e^{-3t} - 2e^{-4t},$$
$$\bar{F}_{3:3}(t) = 2e^{-t} - 2e^{-3t} + e^{-4t},$$
$$\bar{F}_T(t) = 2e^{-3t} - e^{-4t},$$

for all $t \geq 0$. If we assume that $\bar{F}_T$ can be written as a mixture of the functions $\bar{F}_{1:3}$, $\bar{F}_{2:3}$ and $\bar{F}_{3:3}$ with some coefficients c_1, c_2 and c_3, we have

$$2e^{-3t} - e^{-4t} = c_1 e^{-4t} + c_2 \left(e^{-2t} + 2e^{-3t} - 2e^{-4t}\right) + c_3 \left(2e^{-t} - 2e^{-3t} + e^{-4t}\right)$$

for all $t \geq 0$. The functions $e^{-\lambda t}$ and $e^{-\mu t}$ are linearly independent for $\lambda \neq \mu$. Therefore, $c_3 = c_2 = 0$ and we conclude that $\bar{F}_T$ cannot be written as a mixture of $\bar{F}_{1:3}$, $\bar{F}_{2:3}$ and $\bar{F}_{3:3}$. In particular, $\bar{F}_T$ is not equal to the mixture obtained neither with the structural signature $\mathbf{s} = (1/3, 2/3, 0)$ given by

$$\bar{F}_a := \frac{1}{3}\bar{F}_{1:3} + \frac{2}{3}\bar{F}_{2:3}$$

nor with that obtained with the probabilistic signature

$$\bar{F}_p := p_1 \bar{F}_{1:3} + p_2 \bar{F}_{2:3},$$

where $p_i = \Pr(T = X_{i:3})$ for $i = 1, 2$. In this example

$$p_1 = \Pr(X_1 < \min(X_2, X_3)),$$

where X_1 and $Y = \min(X_2, X_3)$ are IID. Therefore, $p_1 = p_2 = 1/2$. The plots of $\bar{F}_T$ (black), $\bar{F}_a$ (blue) and $\bar{F}_p$ (red) and the corresponding hazard rate functions can be seen in Fig. 2.5. Note that the reliability functions are different but similar. The code in R to get the plots of the reliability functions is the following:

```
R13<-function(t)  exp(-4*t)
R23<-function(t)  exp(-2*t)+2*exp(-3*t)-2*exp(-4*t)
R33<-function(t)  2*exp(-t)-2*exp(-3*t)+exp(-4*t)
RT<-function(t)  2*exp(-3*t)-exp(-4*t)
Ra<-function(t)  (1/3)*R13(t)+(2/3)*R23(t)
Rp<-function(t)  (1/2)*R13(t)+(1/2)*R23(t)
curve(R23(x),0,3,lty=2,ylab='Reliability',xlab='t',lwd=2)
curve(R13(x),add=T,lty=2,lwd=2)
curve(R33(x),add=T,lty=2,lwd=2)
curve(RT(x),add=T,lwd=2)
curve(Ra(x),add=T,col='blue',lwd=2)
curve(Rp(x),add=T,col='red',lwd=2)
```

The code in R to get the plots of the hazard rate functions is the following:

```
f1<-function(t)  2*exp(-2*t)
f2<-function(t)  exp(-t)
f13<-function(t)  4*exp(-4*t)
f23<-function(t)  2*exp(-2*t)+6*exp(-3*t)-8*exp(-4*t)
f33<-function(t)  2*exp(-t)-6*exp(-3*t)+4*exp(-4*t)
fT<-function(t)  6*exp(-3*t)-4*exp(-4*t)
fa<-function(t)  (1/3)*f13(t)+(2/3)*f23(t)
fp<-function(t)  (1/2)*f13(t)+(1/2)*f23(t)
curve(f23(x)/R23(x),0,3,ylim=c(0,4),lty=2,lwd=2,ylab='HR')
curve(f13(x)/R13(x),add=T,lty=2,lwd=2)
curve(f33(x)/R33(x),add=T,lty=2,lwd=2)
curve(fT(x)/RT(x),add=T,lwd=2)
curve(fa(x)/Ra(x),add=T,col='blue',lwd=2)
curve(fp(x)/Rp(x),add=T,col='red',lwd=2)
```

◄

Note that in the general case, we can define two mixed systems associated to T, the **average system**

$$\bar{F}_a = s_1 \bar{F}_{1:n} + \cdots + s_n \bar{F}_{n:n}$$

obtained with the structural signature and the **projected system**

$$\bar{F}_p = p_1 \bar{F}_{1:n} + \cdots + p_n \bar{F}_{n:n}$$

obtained with the probabilistic signature. Both can be considered as good approximations of $\bar{F}_T$, see Navarro et al. (2010) (the second one is usually better than the first one as it happen in Fig. 2.5). Note that $\bar{F}_a$ is always the reliability function of a mixed system and that so is $\bar{F}_p$ when $p_1 + \cdots + p_n = 1$. Both $\bar{F}_a$ and

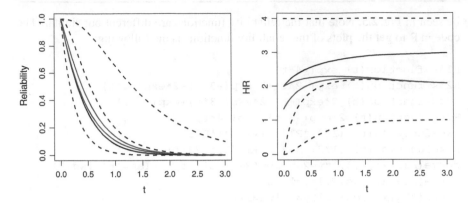

Fig. 2.5 Reliability functions (left) $\bar{F}_T$ (black), $\bar{F}_a$ (blue) and $\bar{F}_p$ (red) and the corresponding hazard rate functions (right) of the system in Example 2.7 when the components are independent with exponential distributions of means $1/2, 1, 1$. The dashed lines represent the functions for the k-out-of-3 systems for $k = 1, 2, 3$

$\bar{F}_b$ belongs to the vectorial space generated by $\bar{F}_{1:n}, \ldots, \bar{F}_{n:n}$. However, this is not always the case for $\bar{F}_T$ as we have seen in Example 2.7.

2.4 Distortion Representations

The distorted distributions were introduced by Wang (1996) and Yaari (1987) in the context of theory of choice under risk. The purpose was to allow a "distortion" (a change) of the initial (or past) risk distribution function. The formal definition is the following.

Definition 2.4 The **distorted distribution** (DD) associated to a distribution function (DF) F and to an increasing and continuous *distortion function* $q : [0, 1] \to [0, 1]$ such that $q(0) = 0$ and $q(1) = 1$, is given by

$$F_q(t) = q(F(t)) \qquad (2.31)$$

for all t.

Note that the conditions on q assure that F_q is a proper distribution function for any distribution function F (actually, for this property, we just need a right-continuous distortion function). Moreover, if q is strictly increasing in $[0, 1]$, then F and F_q have the same support. Also note that q is a distribution function with support included in $[0, 1]$.

From (2.31), we have a similar expression for the respective reliability functions $\bar{F} = 1 - F$ and $\bar{F}_q = 1 - F_q$ that satisfy

$$\bar{F}_q(t) = \bar{q}(\bar{F}(t)), \tag{2.32}$$

where $\bar{q}(u) := 1 - q(1 - u)$ is called the **dual distortion function** in Hürlimann (2004).

Note that $\bar{q}$ is also a "distortion function", that is, it is continuous, increasing and satisfies $\bar{q}(0) = 0$ and $\bar{q}(1) = 1$. Actually, expressions (2.31) and (2.32) are equivalent. However, sometimes it is better to use (2.32) instead of (2.31) (or vice versa).

If F is absolutely continuous with PDF $f = F'$ and q is differentiable, then the PDF of F_q is

$$f_q(t) = f(t)q'(F(t)) = f(t)\bar{q}'(\bar{F}(t)) \tag{2.33}$$

for all t.

From (2.32) and (2.33), the hazard rate function of F_q is

$$h_q(t) = \frac{f_q(t)}{\bar{F}_q(t)} = \frac{\bar{q}'(\bar{F}(t))}{\bar{q}(\bar{F}(t))} f(t) = \alpha(\bar{F}(t))h(t) \tag{2.34}$$

for all t such that $\bar{q}(\bar{F}(t)) > 0$, where $\alpha(u) := u\bar{q}'(u)/\bar{q}(u)$ for $u \in [0, 1]$ and $h(t) = f(t)/\bar{F}(t)$ is the hazard rate function of F.

Analogously, the reversed hazard rate function of F_q is

$$\bar{h}_q(t) = \frac{f_q(t)}{F_q(t)} = \frac{q'(F(t))}{q(F(t))} f(t) = \bar{\alpha}(F(t))\bar{h}(t) \tag{2.35}$$

for all t such that $q(F(t)) > 0$, where $\bar{\alpha}(u) := uq'(u)/q(u)$ for $u \in [0, 1]$ and $\bar{h}(t) = f(t)/F(t)$ is the reversed hazard rate function of F.

However, the expression connecting the MRL functions is not so simple. Thus, if we assume that the support of F is contained in $[0, \infty)$, then

$$m_q(t) = \frac{\int_t^\infty \bar{q}(\bar{F}(x))dx}{\bar{q}(\bar{F}(t))} = \frac{\bar{F}(t)}{\bar{q}(\bar{F}(t))} \frac{\int_t^\infty \bar{q}(\bar{F}(x))dx}{\int_t^\infty \bar{F}(x)dx} m(t)$$

for all t such that $\bar{q}(\bar{F}(t)) > 0$.

Several relevant models are contained in the distorted models. Let us see some of them:

1. **Lehmann's alternatives.** They were introduce in hypothesis testing as an alternative to the distribution function proposed in the null hypothesis. They were defined by

$$F_\theta(t) = F^\theta(t)$$

for all t, where θ is a positive parameter. Clearly, this is a distorted distribution with distortion function $q(u) = u^\theta$ and dual distortion function $\bar{q}(u) = 1 - (1 - u)^\theta$ for $u \in [0, 1]$. The original distribution is obtained with $\theta = 1$.

2. **Proportional hazard rate (PHR) Cox model.** This model was introduced in survival analysis to model the different risks of patients. It is defined by the reliability (survival) function

$$\bar{F}_\theta(t) = \bar{F}^\theta(t)$$

for all t, where θ is a positive parameter. Clearly, this is a distorted distribution with dual distortion function $\bar{q}(u) = u^\theta$ and distortion function $q(u) = 1 - (1 - u)^\theta$. Again, the original distribution is obtained with $\theta = 1$ and, in the absolutely continuous case. Its hazard rate is

$$h_\theta(t) = \theta h(t),$$

that is, the hazard rate function h_θ is proportional to the baseline hazard rate function h. Moreover, the α function in (2.34) is constant and equal to θ. In practice, the θ parameter is obtained (estimated) as $\theta = c_1 x_1 + \cdots + c_k x_k$, where $c_1, \cdots, c_k$ are some positive parameters and $x_1, \ldots, x_k$ represent the characteristics of the patient.

3. **Proportional reversed hazard rate (PRHR) model.** This model is similar to the PHR model and it is defined by the distribution function

$$F_\theta(t) = F^\theta(t)$$

for all t, where θ is a positive parameter. Clearly, this model is equivalent to the Lehmann's alternative model given in item 1 above. In the absolutely continuous case, its reversed hazard rate is

$$\bar{h}_\theta(t) = \theta \bar{h}(t),$$

that is, the reversed hazard rate function is proportional to the baseline reversed hazard rate function (i.e. the function $\bar{\alpha}$ in (2.35) is constant).

4. **Order statistics.** As we have seen in (2.8), the reliability function of the ith order statistic from a sample of IID$\sim F$ random variables can be written as

$$\bar{F}_{i:n}(t) = \sum_{j=0}^{i-1} \binom{n}{j} F^j(t) \bar{F}^{n-j}(t).$$

Hence it is a distorted distribution with dual distortion function

$$\bar{q}_{i:n}(t) = \sum_{j=0}^{i-1} \binom{n}{j} (1-u)^j u^{n-j}.$$

The similar expression for the distortion function is

$$q_{i:n}(t) = \sum_{j=i}^{n} \binom{n}{j} u^j (1-u)^{n-j}.$$

Note that both are polynomials (based on Bernstein polynomials $B_j^n(u) = \binom{n}{j} u^j (1-u)^{n-j}$). Actually, these distortion functions can also be obtained from (2.26) and (2.25) as

$$q_{i:n}(u) = \sum_{j=i}^{n} (-1)^{j-i} \binom{n}{j} \binom{j-1}{i-1} u^j$$

and

$$\bar{q}_{i:n}(u) = \sum_{j=n-i+1}^{n} (-1)^{j-n+i-1} \binom{n}{j} \binom{j-1}{n-i} u^j.$$

Alternatively, we can also use their maximal and minimal signatures, respectively. In particular, for the minimum (series system) and maximum (parallel system) values we have $\bar{F}_{1:n} = \bar{F}^n$ and $F_{n:n} = F^n$. So they are included in the PHR and PRHR models, respectively.

The distorted distributions were generalized in Navarro et al. (2016) as follows.

Definition 2.5 The **distorted distribution** (DD) associated to n distribution functions $F_1, \ldots, F_n$ and to an increasing and continuous **distortion function** $Q : [0, 1]^n \to [0, 1]$ such that $Q(0, \ldots, 0) = 0$ and $Q(1, \ldots, 1) = 1$, is given by

$$F_Q(t) = Q(F_1(t), \ldots, F_n(t)) \tag{2.36}$$

for all t.

As above, the conditions on Q assure that F_Q is a proper distribution function for any distribution functions $F_1, \ldots, F_n$ (actually, for this property, we just need a right-continuous distortion function). Moreover, from (2.36), we have a similar expression for the respective reliability functions

$$\bar{F}_Q(t) = \bar{Q}(\bar{F}_1(t), \ldots, \bar{F}_n(t)), \tag{2.37}$$

where $\bar{Q}(u_1, \ldots, u_n) := 1 - Q(1 - u_1, \ldots, 1 - u_n)$ is called the **dual distortion function**. Note that $\bar{Q}$ is also a "distortion function", that is, it is continuous, increasing and satisfies $\bar{Q}(0, \ldots, 0) = 0$ and $\bar{Q}(1, \ldots, 1) = 1$. Actually, expressions (2.36) and (2.37) are equivalent. However, sometimes it could be better to use (2.37) instead of (2.36) (or vice versa). Note that these expressions are similar to copula representations but that here we obtain a univariate distribution (or reliability) function. The distortion functions are continuous aggregation functions (see Grabisch et al. 2009).

If $F_1, \ldots, F_n$ are absolutely continuous with probability density functions $f_1, \ldots, f_n$ and Q is differentiable, then the PDF of F_Q is

$$f_Q(t) = \sum_{i=1}^{n} f_i(t)\, \partial_i Q(F_1(t), \ldots, F_n(t)) = \sum_{i=1}^{n} f_i(t)\, \partial_i \bar{Q}(\bar{F}_1(t), \ldots, \bar{F}_n(t)), \tag{2.38}$$

for all t, where $\partial_i G$ represents the partial derivative of G with respect to its ith variable.

From (2.37) and (2.38), the hazard rate function of F_q is

$$h_Q(t) = \frac{\sum_{i=1}^{n} f_i(t)\partial_i \bar{Q}(\bar{F}_1(t), \ldots, \bar{F}_n(t))}{\bar{Q}(\bar{F}_1(t), \ldots, \bar{F}_n(t))} = \sum_{i=1}^{n} \alpha_i(\bar{F}_1(t), \ldots, \bar{F}_n(t))h_i(t) \tag{2.39}$$

for all t such that $\bar{F}_Q(t) > 0$, where

$$\alpha_i(u_1, \ldots, u_n) := \frac{u_i \partial_i \bar{Q}(u_1, \ldots, u_n)}{\bar{Q}(u_1, \ldots, u_n)} \geq 0$$

for $u_1, \ldots, u_n \in [0, 1]$ such that $\bar{Q}(u_1, \ldots, u_n) > 0$ and $h_i(t) = f_i(t)/\bar{F}_i(t)$ for $i = 1, \ldots, n$. A similar expression can be obtained for the reversed hazard rate function.

Let us see some examples.

1. **Finite mixtures.** As we have mentioned above, the distribution function of a finite mixture can be written as

$$F(t) = p_1 F_1(t) + \cdots + p_n F_n(t)$$

for all t, where $p_i \geq 0$ and $p_1 + \cdots + p_n = 1$. Therefore it is a distorted distribution with distortion functions

$$Q(u_1, \ldots, u_n) = \bar{Q}(u_1, \ldots, u_n) = p_1 u_1 + \cdots + p_n u_n.$$

However, note that the negative mixtures cannot be represented as distorted distributions.

2. **Generalized proportional hazard rate (GPHR) model.** The PHR model defined above can be extended by

$$\bar{F}(t) = \bar{F}_1^{\theta_1}(t) \ldots \bar{F}_n^{\theta_n}(t),$$

where $\theta_1, \ldots, \theta_n > 0$. Clearly, this is a distorted distribution with dual distortion function

$$\bar{Q}(u_1, \ldots, u_n) = u_1^{\theta_1} \ldots u_n^{\theta_n}.$$

When $\theta_1 = \cdots = \theta_n = 1$, we obtain the reliability of the series system with n independent components.

3. **Generalized proportional reversed hazard rate (GPRHR) model.** Analogously, the PRHR model defined above can be extended by

$$F(t) = F_1^{\theta_1}(t) \ldots F_n^{\theta_n}(t),$$

where $\theta_1, \ldots, \theta_n > 0$. Clearly, this is a distorted distribution with distortion function

$$Q(u_1, \ldots, u_n) = u_1^{\theta_1} \ldots u_n^{\theta_n}.$$

When $\theta_1 = \cdots = \theta_n = 1$, we obtain the distribution of the parallel system with n independent components.

4. **Aggregation functions.** The continuous aggregation functions are equivalent to distorted distributions. So we can use them to get new (distorted) distributions. For example, we can use the arithmetic mean

$$\bar{u} = A_1(u_1, \ldots, u_n) := \frac{u_1 + \cdots + u_n}{n}.$$

This is also a mixture model (with a uniform mixing distribution). Another example is the geometric mean

$$u_g = A_2(u_1, \ldots, u_n) := \sqrt[n]{u_1 \ldots u_n}.$$

If it is applied to the reliability functions, then it is included in the GPHR model and if it is applied to the distribution functions, then it is included in the GPRHR model.

The goal of this section is to prove that the distribution function of a system can be written as a distortion of the distribution functions of the components. To this end we will use the copula theory. The main properties of copulas can be seen in Nelsen (2006) and Durante and Sempi (2016). Thus, if the random vector $\mathbf{X} = (X_1, \ldots, X_n)$ contains the lifetimes of the components in a system then, from Sklar's theorem, we know that the joint distribution function $\mathbf{F}$ of $\mathbf{X}$ can be written as

$$\mathbf{F}(x_1, \ldots, x_n) = C(F_1(x_1), \ldots, F_n(x_n)) \tag{2.40}$$

for all $x_1, \ldots, x_n$, where $F_1, \ldots, F_n$ are the marginal (component) distribution functions and C is a **copula function**, that is, it is a distribution function with uniform marginals over the interval $[0, 1]$. Many authors prefer to restrict copula functions to $C : [0, 1]^n \rightarrow [0, 1]$. In this case, they can always be extended to determine an n-dimensional distribution function with uniform marginals. Moreover, if all the marginal distribution functions $F_1, \ldots, F_n$ are continuous, then C is unique. We also know that if C is a copula, then the right hand side of (2.40) determines a proper joint distribution function for all univariate distribution functions $F_1, \ldots, F_n$ (i.e., from a copula C, we can construct multivariate models with a fixed dependence structure and arbitrary univariate marginals).

A similar representation holds for the reliability functions, that is, the joint reliability function satisfies

$$\bar{\mathbf{F}}(x_1, \ldots, x_n) = \widehat{C}(\bar{F}_1(x_1), \ldots, \bar{F}_n(x_n)) \tag{2.41}$$

for all $x_1, \ldots, x_n$, where $\bar{F}_1, \ldots, \bar{F}_n$ are the marginal (component) reliability functions and $\widehat{C}$ is a copula function, called **survival copula**. It is easy to see that C determines $\widehat{C}$ and vice versa. For example, if $n = 2$, then

$$\widehat{C}(u_1, u_2) = u_1 + u_2 - 1 + C(1 - u_1, 1 - u_2)$$

for all $u_1, u_2 \in [0, 1]$.

We can use (2.41) to prove that the series systems have distorted distributions. For example, if we consider $X_{1:n} = \min(X_1, \ldots, X_n)$, then its reliability function is

$$\bar{F}_{1:n}(t) = \Pr(X_{1:n} > t) = \Pr(X_1 > t, \ldots, X_n > t) = \widehat{C}(\bar{F}_1(t), \ldots, \bar{F}_n(t)),$$

that is, it is a distorted distribution with dual distortion $\bar{Q} = \widehat{C}$. Note that copula functions satisfy the properties of distorted functions but that the reverse is not true (we will see an example later).

If we consider the series system with just the first k components for $k = 1, \ldots, n$, then its lifetime is $X_{1:k} = \min(X_1, \ldots, X_k)$ and its reliability function is

$$\bar{F}_{1:k}(t) = \Pr(X_{1:k} > t) = \Pr(X_1 > t, \ldots, X_k > t) = \widehat{C}(\bar{F}_1(t), \ldots, \bar{F}_k(t), 1, \ldots, 1)$$

that is, it is a distorted distribution with dual distortion function

$$\bar{Q}(u_1, \ldots, u_n) = \widehat{C}(u_1, \ldots, u_k, 1, \ldots, 1)$$

for $u_1, \ldots, u_n \in [0, 1]$.

In the general case, if we consider the series system formed with the components in the set $P \subseteq [n]$, then its lifetime is $X_P = \min_{j \in P} X_j$ and its reliability function is

$$\bar{F}_P(t) = \Pr(X_P > t) = \Pr(\cap_{j \in P}\{X_i > t\}) = \widehat{C}_P(\bar{F}_1(t), \ldots, \bar{F}_n(t)), \qquad (2.42)$$

where

$$\widehat{C}_P(u_1, \ldots, u_n) := \widehat{C}(u_1^P, \ldots, u_n^P), \qquad (2.43)$$

$u_i^P = u_i$ if $i \in P$ and $u_i^P = 1$ if $i \notin P$ for $u_1, \ldots, u_n \in [0, 1]$. Hence all the series systems have distorted distributions. Similar representations can be proved for the parallel systems by using (2.40).

Now we are in a position to prove the main result of this section which says that the same property holds for any semi-coherent system.

Theorem 2.9 (Distortion representation, general case) *If T is the lifetime of a semi-coherent system and the component lifetimes $(X_1, \ldots, X_n)$ have the survival copula $\widehat{C}$, then the reliability function of T can be written as*

$$\bar{F}_T(t) = \bar{Q}(\bar{F}_1(t), \ldots, \bar{F}_n(t)) \qquad (2.44)$$

for all t, where $\bar{Q}$ is a distortion function which depends on ψ and $\widehat{C}$.

Proof From the minimal path set representation (2.15), we have

$$\bar{F}_T(t) = \sum_{i=1}^{r} \bar{F}_{P_i}(t) - \sum_{i=1}^{r-1} \sum_{j=i+1}^{r} \bar{F}_{P_i \cup P_j}(t) + \cdots + (-1)^{r+1} \bar{F}_{P_1 \cup \cdots \cup P_r}(t).$$

Hence, from (2.42) and (2.43), we obtain (2.44) with

$$\bar{Q}(\mathbf{u}) = \sum_{i=1}^{r} \widehat{C}_{P_i}(\mathbf{u}) - \sum_{i=1}^{r-1} \sum_{j=i+1}^{r} \widehat{C}_{P_i \cup P_j}(\mathbf{u}) + \cdots + (-1)^{r+1} \widehat{C}_{P_1 \cup \cdots \cup P_n}(\mathbf{u}) \quad (2.45)$$

for $\mathbf{u} = (u_1, \ldots, u_n) \in [0, 1]^n$, where $\widehat{C}_P$ is defined by (2.43). The function $\bar{Q}$ is always a distortion function since $\bar{F}_T$ is a proper reliability function for all $\bar{F}_1, \ldots, \bar{F}_n$. $\qquad \square$

A similar proof can be obtained by using parallel systems and the minimal cut set representation. The function $\bar{Q}$ can be called **distortion function (or domination) of the system**. Note that it depends on both the structure (the minimal path sets) of the systems and on the structure dependence between the component lifetimes (the survival copula). However, it does not depend on the component (marginal) reliability functions. So (2.44) is a very convenient representation for the system reliability since all the system characteristics (dependence and structure) are included $\bar{Q}$ and the different units are represented by their different marginal reliability functions. In many situations in practice, we can choose different units (reliabilities) for a fixed system structure Q or study different system characteristics (different $\bar{Q}$ functions) for arbitrary or fixed components.

Next we analyse some particular cases of interest.

Theorem 2.10 (Distortion representation, IND case) *If T is the lifetime of a semi-coherent system with independent component lifetimes $X_1, \ldots, X_n$, then the reliability function of T can be written as*

$$\bar{F}_T(t) = \bar{Q}(\bar{F}_1(t), \ldots, \bar{F}_n(t))$$

for all t, where $\bar{Q}$ is a multinomial which only depends on ψ.

The proof is immediate from (2.17) or (2.45). The multinomial $\bar{Q}$ was called **reliability function of the structure** ψ in Barlow and Proschan (1975), p. 21. However note that, $\bar{Q}$ is not a joint reliability function (it is a distortion function). Also note that this multinomial can be obtained by using the product-coproduct representations for the structure function given in (1.6) and (1.7). It can also be obtained from the pivotal decomposition (1.3) or from the Möbius representation (1.10).

In the general case, the distortion function $\bar{Q}$ in (2.45) can also be obtained from the Möbius transform $\hat{\varphi}$ and $\hat{C}$ as

$$\bar{Q}(\mathbf{u}) = \sum_{I \subseteq [n]} \hat{\varphi}(I)\hat{C}(\mathbf{u}_I),$$

where $\mathbf{u} = (u_1, \ldots, u_n)$ and $\mathbf{u}_I = (u_1^I, \ldots, u_n^I)$ with $u_i^I = u_i$ if $i \in I$ and $u_i^I = 1$ if $i \notin I$, see (3.6) in Navarro and Spizzichino (2020).

Theorem 2.11 (Distortion representation, ID case) *If T is the lifetime of a semi-coherent system and the component lifetimes $(X_1, \ldots, X_n)$ have the survival copula $\hat{C}$ and a common reliability $\bar{F}$, then the reliability function of T can be written as*

$$\bar{F}_T(t) = \bar{q}(\bar{F}(t))$$

for all t, where $\bar{q}$ is a distortion function which only depends on ψ and on $\hat{C}$.

The proof is immediate from (2.44) with

$$\bar{q}(u) = \bar{Q}(u, \ldots, u)$$

for $u \in [0, 1]$. In particular, in the EXC case, $\bar{q}$ can be written as

$$\bar{q}(u) = \sum_{i=1}^{n} a_i \widehat{C}(\underbrace{u, \ldots, u}_{i \; times}, \underbrace{1, \ldots, 1}_{n-i \; times}),$$

where $(a_1, \ldots, a_n)$ is the minimal signature of order n.

Theorem 2.12 (Distortion representation, IID case) *If T is the lifetime of a semi-coherent system with IID component lifetimes $X_1, \ldots, X_n$ having a common reliability $\bar{F}$, then the reliability function of T can be written as*

$$\bar{F}_T(t) = \bar{q}(\bar{F}(t))$$

for all t, where $\bar{q}(u) = \sum_{i=1}^{n} a_i u^i$ is a distortion function and $(a_1, \ldots, a_n)$ is the minimal signature of order n.

The proof is immediate from the two preceding theorems or from the minimal signature representation given in (2.22). Note that, in this case, $\bar{q}$ is the polynomial obtained with the minimal signature coefficients.

Let us see some examples. The simplest one is the representation of the components. Thus, the reliability function of X_i can be written as

$$\bar{F}_i(t) = \bar{Q}_i(\bar{F}_1(t), \ldots, \bar{F}_n(t))$$

for $\bar{Q}_i(u_1, \ldots, u_n) = u_i$ and $i = 1, \ldots, n$.

As mentioned above, the representation for the series systems is also immediate. In particular, the reliability function of $X_{1:k}$ is

$$\bar{F}_{1:k}(t) = \bar{Q}_{1:k}(\bar{F}_1(t), \ldots, \bar{F}_n(t))$$

for

$$\bar{Q}_{1:k}(u_1, \ldots, u_n) = \widehat{C}(u_1, \ldots, u_k, 1, \ldots, 1)$$

for $k = 1, \ldots, n$. If the components are IND, then

$$\bar{Q}_{1:k}(u_1, \ldots, u_n) = u_1 \ldots u_k.$$

If the components are ID with a common reliability $\bar{F}$, then $\bar{F}_{1:k}(t) = \bar{q}_{1:k}(\bar{F}(t))$ with

$$\bar{q}_{1:k}(u) = \bar{Q}_{1:k}(u, \ldots, u) = \widehat{C}(\underbrace{u, \ldots, u}_{k \; times}, \underbrace{1, \ldots, 1}_{n-k \; times})$$

and, in particular, it they are IID, then $\bar{q}_{1:k}(u) = u^k$ for $k = 1, \ldots, n$.

For the parallel systems, it is better to use the distributional copula C. Thus the distribution function of $X_{k:k}$ can be written as

$$F_{k:k}(t) = Q_{k:k}(F_1(t), \ldots, F_n(t))$$

for

$$Q_{k:k}(u_1, \ldots, u_n) = C(u_1, \ldots, u_k, 1, \ldots, 1)$$

for $k = 1, \ldots, n$. Hence, its reliability function is

$$\bar{F}_{k:k}(t) = \bar{Q}_{k:k}(\bar{F}_1(t), \ldots, \bar{F}_n(t))$$

for

$$\bar{Q}_{k:k}(u_1, \ldots, u_n) = 1 - C(1 - u_1, \ldots, 1 - u_k, 1, \ldots, 1)$$

for $k = 1, \ldots, n$.

We can also obtain representations based on $\widehat{C}$ from the minimal path set representation. For example, for $X_{2:2}$ we get

$$\bar{F}_{2:2}(t) = \bar{F}_1(t) + \bar{F}_2(t) - \bar{F}_{1:2}(t) = \bar{Q}_{2:2}(\bar{F}_1(t), \ldots, \bar{F}_n(t))$$

with

$$\bar{Q}_{2:2}(u_1, \ldots, u_n) = u_1 + u_2 - \widehat{C}(u_1, u_2).$$

A similar expression can be obtained for $X_{k:k}$. If the components are IND, then

$$\bar{Q}_{k:k}(u_1, \ldots, u_n) = 1 - (1 - u_1) \ldots (1 - u_k) = \coprod_{i=1}^{k} u_i.$$

If they are ID, then $\bar{F}_{k:k}(t) = \bar{q}_{k:k}(\bar{F}(t))$ for

$$\bar{q}_{k:k}(u) = 1 - C(\underbrace{1 - u, \ldots, 1 - u}_{k\ times}, \underbrace{1, \ldots, 1}_{n-k\ times})$$

and, if they are IID, then $\bar{q}_{k:k}(u) = 1 - (1 - u)^k$.

We can also consider a general coherent system. For example, for our favourite system $T = \min(X_1, \max(X_2, X_3))$, we have

$$\bar{F}_T(t) = \bar{F}_{\{1,2\}}(t) + \bar{F}_{\{1,3\}}(t) - \bar{F}_{1:3}(t) = \bar{Q}_T(\bar{F}_1(t), \bar{F}_2(t), \bar{F}_3(t))$$

with

$$\bar{Q}_T(u_1, u_2, u_3) = \widehat{C}(u_1, u_2, 1) + \widehat{C}(u_1, 1, u_3) - \widehat{C}(u_1, u_2, u_3).$$

If the components are IND, then

$$\bar{Q}_T(u_1, u_2, u_3) = u_1 u_2 + u_1 u_3 - u_1 u_2 u_3 = u_1 (u_2 \coprod u_3).$$

If they are ID, then $\bar{F}_T(t) = \bar{q}_T(\bar{F}(t))$ with

$$\bar{q}_T(u) = \widehat{C}(u, u, 1) + \widehat{C}(u, 1, u) - \widehat{C}(u, u, u)$$

and, if they are IID, then $\bar{q}_T(u) = 2u^2 - u^3$ for $u \in [0, 1]$. Recall that its minimal signature is $(0, 2, -1)$.

Proceeding in a similar way we can obtain the dual distortion functions given in Tables 2.4 and 2.5 for all the systems with 1-3 IND and IID components, respectively. In the second case all the systems equivalent under permutations have the same distortions (so they are not repeated in the table).

Table 2.4 Dual distortions functions for all the systems with 1-3 IND components

	$T = \psi(X_1, X_2, X_3)$	$\bar{Q}(u_1, u_2, u_3)$
1	$X_{1:3} = \min(X_1, X_2, X_3)$	$u_1 u_2 u_3$
2	$\min(X_2, X_3)$	$u_2 u_3$
3	$\min(X_1, X_3)$	$u_1 u_3$
4	$\min(X_1, X_2)$	$u_1 u_2$
5	$\min(X_3, \max(X_1, X_2))$	$u_1 u_3 + u_2 u_3 - u_1 u_2 u_3$
6	$\min(X_2, \max(X_1, X_3))$	$u_1 u_2 + u_2 u_3 - u_1 u_2 u_3$
7.	$\min(X_1, \max(X_2, X_3))$	$u_1 u_2 + u_1 u_3 - u_1 u_2 u_3$
8	X_3	u_3
9	X_2	u_2
10	X_1	u_1
11	$X_{2:3}$	$u_1 u_2 + u_1 u_3 + u_2 u_3 - 2u_1 u_2 u_3$
12	$\max(X_3, \min(X_1, X_2))$	$u_3 + u_1 u_2 - u_1 u_2 u_3$
13	$\max(X_2, \min(X_1, X_3))$	$u_2 + u_1 u_3 - u_1 u_2 u_3$
14	$\max(X_1, \min(X_2, X_3))$	$u_1 + u_2 u_3 - u_1 u_2 u_3$
15	$\max(X_2, X_3)$	$u_2 + u_3 - u_2 u_3$
16	$\max(X_1, X_3)$	$u_1 + u_3 - u_1 u_3$
17	$\max(X_1, X_2)$	$u_1 + u_2 - u_1 u_2$
18	$X_{3:3} = \max(X_1, X_2, X_3)$	$u_1 + u_2 + u_3 - u_1 u_2 - u_1 u_3 - u_2 u_3 + u_1 u_2 u_3$

Table 2.5 Dual distortions functions for all the systems with 1-3 IID components

	$T = \psi(X_1, X_2, X_3)$	$\bar{q}(u)$
1	$X_{1:3} = \min(X_1, X_2, X_3)$	u^3
2	$X_{1:2} = \min(X_1, X_2)$	u^2
3	$\min(X_1, \max(X_2, X_3))$	$2u^2 - u^3$
4	X_1	u
5	$X_{2:3}$	$3u^2 - 2u^3$
6	$\max(X_1, \min(X_2, X_3))$	$u + 2u^2 - u^3$
7	$X_{2:2} = \max(X_1, X_2)$	$2u - u^2$
8	$X_{3:3} = \max(X_1, X_2, X_3)$	$3u - 3u^2 + u^3$

The preceding representations can be used jointly with the representation based on distortions to compute the reliability and hazard rate functions of a system. For example, in Fig. 2.6, we plot the reliability functions for series and parallel systems of order 2 when the component lifetimes have exponential distributions of means 1 and 1/2 and when they are independent (left) or they have the following Clayton–Oakes survival copula (right)

$$\widehat{C}(u, v) = \frac{uv}{u + v - uv}$$

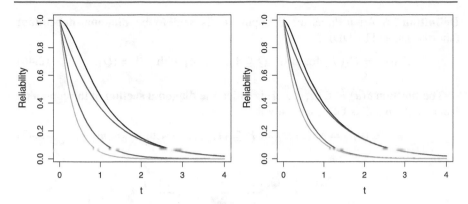

Fig. 2.6 Reliability functions for the parallel system $X_{2:2}$ (black) and the series system $X_{1:2}$ (green) when the components have exponential distributions of means 1 (red) and 1/2 (blue) and they are independent (left) or dependent with a Clayton–Oakes copula (right)

for $u, v \in [0, 1]$. This copula induces a positive dependence between the component lifetimes. Note that, in both cases, $\bar{F}_{1:2} \leq \bar{F}_i \leq \bar{F}_{2:2}$ holds (this property is always true) and that the positive dependence induced by this copula improves the series system but that it gets worse the parallel system (as expected). The code in R to get the right plot is the following. By changing C we can obtain other plots (including the left plot).

```
C<-function(u,v) u*v/(u+v-u*v)
Q<-function(u,v) u+v-C(u,v)
R1<-function(t) exp(-t)
R2<-function(t) exp(-2*t)
R<-function(t) Q(R1(t),R2(t))
curve(R(x),0,4,xlab='t',ylab='Reliability',lwd=2)
curve(R1(x),add=T,col='red',lwd=2)
curve(R2(x),add=T,col='blue',lwd=2)
curve(C(R1(x),R2(x)),add=T,col='green',lwd=2)
```

We conclude this section by extending the signature representations. We have seen in the preceding section that they hold when the component lifetimes have an exchangeable (EXC) joint distribution function $\mathbf{F}$. This condition is equivalent to have ID components and an EXC survival copula $\widehat{C}$. We have also proved above that the ID condition cannot be dropped-out. However, let us see that the condition: "$\widehat{C}$ is EXC", can be relaxed. To this end we need the following concept extracted from Okolewski (2017). Recall that we use the following notation. For any set $I \subseteq \{1, \ldots, n\}$, $\mathbf{u}_I := (u_1, \ldots, u_n)$ denotes the vector with $u_i = u$ for $i \in I$ and $u_i = 1$ if $i \notin I$. The cardinality of the set I is denoted by $|I|$.

Definition 2.6 An n-dimensional copula C is said to be **diagonal-dependent** (shortly denoted by DD) if

$$C(\mathbf{u}_P) = C(\mathbf{u}_Q) \text{ for all } P, Q \subseteq \{1, \ldots, n\} \text{ with } |P| = |Q|. \tag{2.46}$$

The function $\delta(u) = C(u, \ldots, u)$ is called the **diagonal section** of the copula C. Hence note that C is DD if and only if

$$C(\mathbf{u}_P) = \delta_m(u) \text{ for all } P \subseteq \{1, \ldots, n\} \text{ with } |P| = m \tag{2.47}$$

for $m = 1, \ldots, n$, where

$$\delta_m(u) := C(\underbrace{u, \ldots, u,}_{m-times} \underbrace{1, \ldots, 1}_{(n-m)-times})$$

is the diagonal section for the copula of the marginal distribution of the first m-variables. Clearly, $\delta_n(u) = C(u, \ldots, u) = \delta(u)$ and $\delta_1(u) = u$ for all $u \in [0, 1]$ (since all the univariate marginals have a uniform distribution over the interval $(0, 1)$). So we just need to check (2.47) for $m = 2, \ldots, n - 1$.

In particular, a copula C is DD when all the marginals of dimension m have the same copula for all $1 < m < n$. Of course, all the EXC copulas are, in particular, DD. The reverse is not true, see the counterexample given in Navarro and Fernández-Sánchez (2020).

Now we are ready to state the following result extracted from Navarro and Fernández-Sánchez (2020).

Theorem 2.13 (Distortion reprersentation, DD-ID case) *If T is the lifetime of a semi-coherent system and the component lifetimes $(X_1, \ldots, X_n)$ are ID and have a DD survival copula, then (2.5) holds for the structural signature of dimension n.*

Proof From (2.15) we know that the system reliability function $\bar{F}_T$ can be written as a linear combination of the reliability functions of the series systems. If the component lifetimes are ID with a common reliability function $\bar{F}$ and a DD survival copula $\hat{C}$, then

$$\bar{F}_P(t) = \mathbb{P}\left(\min_{j \in P} T_j > t\right) = \hat{C}_P(\bar{F}(t), \ldots, \bar{F}(t)) = \hat{\delta}_m(\bar{F}(t)) \tag{2.48}$$

holds for all t and all $P \subseteq \{1, \ldots, n\}$ with $|P| = m$, where

$$\hat{\delta}_m(u) := \hat{C}(\underbrace{u, \ldots, u,}_{m-times} \underbrace{1, \ldots, 1}_{(n-m)-times})$$

for all $u \in [0, 1]$ and $m = 1, \ldots, n$. Hence, all the series systems with the same number of components m do have the same reliability function given by (2.48). Therefore, the general representation (2.15) can be reduced to

$$\bar{F}_T(t) = a_1 \hat{\delta}_1(\bar{F}(t)) + \cdots + a_n \hat{\delta}_n(\bar{F}(t)), \tag{2.49}$$

where $\mathbf{a} = (a_1, \ldots, a_n)$ is the minimal signature of order n of the system.

The preceding representation (2.49) holds for any system structure (with the appropriate coefficients $a_1, \ldots, a_n$). For example, the series system with n components has just one minimal path set $P_1 = \{1, \ldots, n\}$ and lifetime $T_{1:n} = \min(T_1, \ldots, T_n)$. Hence

$$\bar{F}_{1:n}(t) = \Pr(T_1 > t, \ldots, T_n > t) = \widehat{C}(\bar{F}(t), \ldots, \bar{F}(t)) = \widehat{\delta}_n(\bar{F}(t))$$

for all t.

Analogously, the minimal path sets of $T_{2:n}$ are all the subsets with $n-1$ elements. So it has $n = \binom{n}{n-1}$ minimal path sets and, from (2.15),

$$\bar{F}_{2:n}(t) = n\widehat{\delta}_{n-1}(\bar{F}(t)) - (n-1)\widehat{\delta}_n(\bar{F}(t))$$

holds for all t. The last coefficient in the preceding expression is $n-1$ because the coefficients in (2.49) sum up to 1 (take $t \to -\infty$).

In general, $T_{i:n}$ has $\binom{n}{n-i+1}$ minimal path sets and, from (2.15), its reliability function can be written as

$$\bar{F}_{i:n}(t) = a_{i,n-i+1}\widehat{\delta}_{n-i+1}(\bar{F}(t)) + \cdots + a_{i,n}\widehat{\delta}_n(\bar{F}(t)) \qquad (2.50)$$

for some coefficients $a_{i,n-i+1}, \ldots, a_{i,n}$ such that $a_{i,n-i+1} + \cdots + a_{i,n} = 1$ and $a_{i,n-i+1} = \binom{n}{n-i+1}$ for $i = 1, \ldots, n$.

Thus, if we define the column vectors $\mathbf{r}(t) = (\bar{F}_{1:n}(t), \ldots, \bar{F}_{n:n}(t))'$ and $\mathbf{d}(t) = (\widehat{\delta}_1(\bar{F}(t)), \ldots, \widehat{\delta}_n(\bar{F}(t)))'$, (2.50) proves that $\mathbf{r}(t) = A_n \mathbf{d}(t)$ for a triangular real-valued matrix $A_n = (a_{i,j})$ such that $a_{i,n-i+1} = \binom{n}{n-i+1}$ and $a_{i,j} = 0$ for $i = 1, \ldots, n$ and $j = 1, \ldots, n-i$. Hence A_n is not singular and so we can write $\mathbf{d}(t) = A_n^{-1}\mathbf{r}(t)$, where A_n^{-1} is the inverse matrix of A_n. Moreover, note that (2.49) can be rewritten as $\bar{F}_T(t) = \mathbf{a}\,\mathbf{d}(t)$. Then

$$\bar{F}_T(t) = \mathbf{a}A_n^{-1}\mathbf{r}(t) = (c_1, \ldots, c_n)\mathbf{r}(t) = c_1\bar{F}_{1:n}(t) + \cdots + c_n\bar{F}_{n:n}(t)$$

for all t, where $(c_1, \ldots, c_n) := \mathbf{a}A_n^{-1}$ are some coefficients which only depend on the structure of the system. Therefore, these coefficients should be the same as that obtained in the IID continuous case, that is, $c_i = s_i^{(n)}$ for $i = 1, \ldots, n$. So (2.5) holds with the same coefficients for systems with ID component lifetimes and DD survival copulas. $\square$

In Navarro and Fernández-Sánchez (2020) it is proved that the set $\mathcal{S}_{DD}$ of all the DD copulas is much bigger than the set $\mathcal{S}_{EXC}$ of EXC copulas. Actually, $\mathcal{S}_{DD}$ is dense in the set of all the copulas while $\mathcal{S}_{EXC}$ is not. Therefore, for any copula C we can find a "close" DD copula. The following example illustrate these representations.

Example 2.8 Let us consider again $T = \min(X_1, \max(X_2, X_3))$ with

$$\bar{F}(t) = \Pr(X_1 > t, X_2 > t) + \Pr(X_1 > t, X_3 > t) - \Pr(X_1 > t, X_2 > t, X_3 > t).$$

Let us assume

$$\Pr(X_1 > x_1, X_2 > x_2, X_3 > x_3) = \widehat{C}(\bar{F}_1(x_1), \bar{F}_2(x_2), \bar{F}_3(x_3)),$$

where $\widehat{C}$ is the survival copula. If we assume $\bar{F}_1 = \bar{F}_2 = \bar{F}_3 = \bar{F}$ (ID), then

$$\Pr(X_1 > t, X_2 > t) = \widehat{C}(\bar{F}(t), \bar{F}(t), 1)$$
$$\Pr(X_1 > t, X_3 > t) = \widehat{C}(\bar{F}(t), 1, \bar{F}(t))$$
$$\Pr(X_1 > t, X_2 > t, X_3 > t) = \widehat{C}(\bar{F}(t), \bar{F}(t), \bar{F}(t)).$$

Therefore, $\bar{F}_T(t) = \bar{q}(\bar{F}(t))$ with

$$\bar{q}(u) = \widehat{C}(u, u, 1) + \widehat{C}(u, 1, u) - \widehat{C}(u, u, u).$$

Analogously, it can be proved that $\bar{F}_{i:3}(t) = \bar{q}_{i:3}(\bar{F}(t))$ for $i = 1, 2$ with

$$\bar{q}_{1:3}(u) = \widehat{C}(u, u, u)$$
$$\bar{q}_{2:3}(u) = \widehat{C}(u, u, 1) + \widehat{C}(u, 1, u) + \widehat{C}(1, u, u) - 2\widehat{C}(u, u, u).$$

As the structural signature is $s = (1/3, 2/3, 0)$, we do not need $\bar{F}_{3:3}$.

If the components are IID, that is, $\widehat{C}(u_1, u_2, u_3) = u_1 u_2 u_3$, then

$$\bar{q}(u) = 2u^2 - u^3$$
$$\bar{q}_{1:3}(u) = u^3$$
$$\bar{q}_{2:3}(u) = 3u^2 - 2u^3.$$

Therefore

$$\bar{q}(u) = \frac{1}{3}\bar{q}_{1:3}(u) + \frac{2}{3}\bar{q}_{1:3}(u)$$

holds since

$$2u^2 - u^3 = \frac{1}{3}(u^3) + \frac{2}{3}(3u^2 - 2u^3).$$

If $\widehat{C}$ is DD, then

$$\widehat{C}(u, u, 1) = \widehat{C}(u, 1, u) = \widehat{C}(1, u, u)$$

and so

$$\bar{q}(u) = 2\widehat{C}(u, u, 1) - \widehat{C}(u, u, u)$$
$$\bar{q}_{1:3}(u) = \widehat{C}(u, u, u)$$
$$\bar{q}_{2:3}(u) = 3\widehat{C}(u, u, 1) - 2\widehat{C}(u, u, u)$$

for all $u \in [0, 1]$. Therefore

$$\bar{q}(u) = \frac{1}{3}\bar{q}_{1:3}(u) + \frac{2}{3}\bar{q}_{1:3}(u)$$

holds since

$$2\widehat{C}(u, u, 1) - \widehat{C}(u, u, u) = \frac{1}{3}\widehat{C}(u, u, u) + \frac{2}{3}(3\widehat{C}(u, u, 1) - 2\widehat{C}(u, u, u)).$$

However, if $\widehat{C}$ is the following Farlie-Gumbel-Morgenstern (FGM) copula:

$$\widehat{C}(u_1, u_2, u_3) = u_1 u_2 u_3(1 + \theta(1 - u_2)(1 - u_3))$$

for $-1 \le \theta \le 1$, then

$$\bar{q}(u) = 2u^2 - \widehat{C}(u, u, u)$$
$$\bar{q}_{1:3}(u) = \widehat{C}(u, u, u)$$
$$\bar{q}_{2:3}(u) = 3u^2 + \theta u^2(1 - u)^2 - 2\widehat{C}(u, u, u).$$

Therefore

$$\bar{q}(u) = \frac{1}{3}\bar{q}_{1:3}(u) + \frac{2}{3}\bar{q}_{1:3}(u)$$

does hold for $\theta \ne 0$ since

$$2u^2 - \widehat{C}(u, u, u) \ne \frac{1}{3}\widehat{C}(u, u, u) + \frac{2}{3}(3u^2 + \theta u^2(1 - u)^2 - 2\widehat{C}(u, u, u))$$

for $0 < u < 1$.   ◀

Problems

1. Prove that if X is a non-negative random variable, then

$$E(X) = \int_0^\infty \bar{F}_X(x)dx.$$

2. Compute the MTTF in the exponential model.
3. Prove that the exponential model satisfies the lack of memory property.
4. Prove that the exponential model is the unique continuous model that satisfies the lack of memory property.
5. Prove that the MRL of the exponential model satisfies $m(t) = \mu$ for all $t \ge 0$.
6. Prove that the hazard rate of the exponential model satisfies $h(t) = 1/\mu$ for all $t \ge 0$.
7. Obtain the hazard rate of the Weibull model.
8. Obtain the reliability function of a model with hazard rate $h(t) = a + bt$ for $t \ge 0$ and $a, b \ge 0$.
9. Obtain the reliability function of a model with hazard rate $h(t) = 1/(a + bt)$ for $t \ge 0$ and $a, b \ge 0$.
10. Obtain the relationship between the reversed hazard rate and mean inactivity time functions.
11. Obtain a representation similar to (2.14) for the MRL of the system in the IID continuous case.
12. Obtain the minimal path set representation of a system of order 4.
13. Obtain the minimal cut set representation of a system of order 4.
14. Obtain the minimal signature representation of a system of order 4.
15. Obtain the maximal signature representation of a system of order 4.
16. Compute the matrices A_4 and A_4^{-1}.
17. Compute the matrices B_4 and B_4^{-1}.
18. Compute the matrix C_4.
19. Obtain the signature of order 4 of a coherent system of order 3.
20. Obtain the signature of order 5 of a coherent system of order 4.

21. Prove with an example that Samaniego's representation does not hold without the EXC assumption.
22. Prove that the function in (2.12) is a proper PDF for $i, n \in \mathbb{R}$ satisfying $1 \le i \le n$.
23. Prove that $\widehat{C}(u_1, u_2) = u_1 + u_2 - 1 + C(1 - u_1, 1 - u_2)$.
24. Compute the distortion functions of a system of order 4.
25. Use the distortion function of a system to plot its reliability and hazard rate functions.
26. Compare the reliability functions of two systems by using distortions.
27. Compare the hazard rate functions of two systems by using distortions.
28. Obtain the signature representation for a DD (non-EXC) copula.

Stochastic Comparisons

3

Abstract

In this chapter we use the representations obtained in the preceding chapter to stochastically compare the performance of different systems. We consider the main stochastic orders: the usual stochastic order, the hazard rate order, the mean residual life order, the reversed hazard rate order and the likelihood ratio order. We use different techniques depending on the assumptions made about the components. We consider systems with independent and identically distributed (IID) components, exchangeable (EXC) components, identically distributed (ID) components, independent (IND) components or dependent components. The dependence is modeled by using copulas (or joint reliability functions). This chapter is based on the review paper Navarro (2018b).

3.1 Main Stochastic Orders

First we give the definitions and the main properties of the stochastic orders considered here. Note that they can be used to compare both the system and the component lifetimes (i.e. non-negative random variables). For more properties and applications we refer the interested readers to Belzunce et al. (2016), Müller and Stoyan (2002) and Shaked and Shanthikumar (2007).

If X and Y are two random variables (representing the lifetimes of two different units or systems), there exist several ways to stochastically compare X and Y. The first option is to compare their means (or expected lifetimes) $\mu_X = E(X)$ and $\mu_Y = E(Y)$ (if they exist). Thus we write $X \leq_M Y$ (mean order) when $\mu_X \leq \mu_Y$.

The second main option is the (usual) stochastic order defined as follows.

Definition 3.1 X is said to be smaller than Y in the **stochastic order** (denoted by $X \leq_{ST} Y$ or by $F_X \leq_{ST} F_Y$) if $\bar{F}_X(t) \leq \bar{F}_Y(t)$ for all t, where $\bar{F}_X$ and $\bar{F}_Y$ are the reliability functions of X and Y, respectively.

J. Navarro, *Introduction to System Reliability Theory*,
https://doi.org/10.1007/978-3-030-86953-3_3

Note that here (and throughout the book) 'smaller than' means 'smaller than or equal to'. Also, if both $X \leq_{ST} Y$ and $X \geq_{ST} Y$ hold (i.e., $X =_{ST} Y$), then $\bar{F}_X(t) = \bar{F}_Y(t)$ for all t, that is, they have the same law (distribution). This order can also be called 'the reliability order' since $X \leq_{ST} Y$ means that the reliability of the units represented by Y is always equal to or greater than the reliability of the units represented by X.

The ST order $X \leq_{ST} Y$ is characterized by the following property:

$$E(g(X)) \leq E(g(Y)) \tag{3.1}$$

for any increasing function g such that these expectations exist. This property is sometimes used as a definition. Recall that, throughout the book, we use increasing (decreasing) in the weak sense, that is, a function g is increasing (decreasing) if $g(a) \leq g(b)$ $(\geq)$ for all $a \leq b$. Therefore, the stochastic order can be seen as an extension of the expected value order for increasing functions. In particular, we have that $X \leq_{ST} Y$ implies $E(X) \leq E(Y)$ whenever both expectations exist. Also note that, from (2.3), if $X \leq_{ST} Y$ and $E(X) = E(Y)$ hold, then $X =_{ST} Y$.

Another characterization of this order is the following: $X \leq_{ST} Y$ if and only if there exist two random variables X^* and Y^* over the same probability space (Ω, S, Pr) such that $X^* =_{ST} X$, $Y^* =_{ST} Y$ and $X^*(\omega) \leq Y^*(\omega)$ for all $\omega \in \Omega$ (see Shaked and Shanthikumar 2007, p. 5). However, note that if X and Y are defined over the same probability space Ω, $X \leq_{ST} Y$ does not necessarily imply that $X(\omega) \leq_{ST} Y(\omega)$ for all $\omega \in \Omega$. As an immediate consequence we have that if $X \leq_{ST} Y$, then $aX + b \leq_{ST} aY + b$ for all $a > 0$ and b. The ordering is reversed when $a < 0$.

Another option is to compare X and Y by comparing their respective aging functions. For example, the hazard rate order is defined as follows.

Definition 3.2 X is said to be smaller than Y in the **hazard (or failure) rate** order (denoted by $X \leq_{HR} Y$ or by $F_X \leq_{HR} F_Y$) if $\bar{F}_Y/\bar{F}_X$ is an increasing function (with the convention $a/0 = +\infty$ for all $a > 0$).

The HR order can be characterized in terms of the ST order by the following property:

$$X \leq_{HR} Y \Leftrightarrow (X - t | X > t) \leq_{ST} (Y - t | Y > t) \quad \text{for all } t. \tag{3.2}$$

Hence the HR order can be interpreted as follows: $X \leq_{HR} Y$ if and only if the residual lifetime of a used unit with age t from X is ST-smaller than the residual lifetime of a used unit with the same age t from Y for all t. Note that $X \leq_{HR} Y$ implies $X \leq_{ST} Y$.

If X and Y are two random variables with absolutely continuous (or discrete) distribution functions, then $X \leq_{HR} Y$ iff $h_X(t) \geq h_Y(t)$ for all t, where $h_X = f_X/\bar{F}_X$ and $h_Y = f_Y/\bar{F}_Y$ are the HR functions of X and Y, respectively.

Analogously, the reversed hazard rate order is defined as follows.

Definition 3.3 X is said to be smaller than Y in the **reversed hazard rate order** (denoted by $X \leq_{RHR} Y$ or by $F_X \leq_{RHR} F_Y$) if F_Y/F_X is an increasing function.

The RHR order can be characterized in terms of the ST order by the following property:

$$X \leq_{RHR} Y \Leftrightarrow (X|X \leq t) \leq_{ST} (Y|Y \leq t) \quad \text{for all } t$$

or equivalently, by

$$X \leq_{RHR} Y \Leftrightarrow (t - X|X \leq t) \geq_{ST} (t - Y|Y \leq t) \quad \text{for all } t. \tag{3.3}$$

From (3.3), the RHR order can be interpreted as follows: $X \leq_{RHR} Y$ holds if and only if the inactivity time of a unit which has failed before age t from X is ST-greater than the inactivity time of a unit which has failed before age t from Y for all t.

If X and Y are two random variables with absolutely continuous (or discrete) distribution functions, then $X \leq_{RHR} Y$ iff $\bar{h}_X(t) \leq \bar{h}_Y(t)$ for all t, where $\bar{h}_X = f_X/F_X$ and $\bar{h}_Y = f_Y/F_Y$ are the reverse hazard rate functions of X and Y, respectively.

It can be proved that the RHR order does not imply the HR order and that the HR order does not imply the RHR order. However, they are related by the following properties:

$$X \leq_{RHR} Y \Leftrightarrow -X \geq_{HR} -Y$$

and

$$X \leq_{HR} Y \Leftrightarrow -X \geq_{RHR} -Y$$

since $F_X(t) = \bar{F}_{-X}(-t)$, $f_X(t) = f_{-X}(-t)$, $h_X(t) = \bar{h}_{-X}(-t)$ and $\bar{h}_X(t) = h_{-X}(-t)$.

Next we give the definition of a stronger order also related with conditional expectations and aging properties, the likelihood ratio order.

Definition 3.4 If X and Y are two random variables with absolutely continuous (or discrete) distribution functions, X is said to be smaller than Y in the **likelihood ratio order** (denoted by $X \leq_{LR} Y$ or by $F_X \leq_{LR} F_Y$) if f_Y/f_X is increasing in the union of their supports, where f_X and f_Y are probability density (or probability mass) functions of X and Y, respectively.

Note that $X \leq_{LR} Y$ holds if and only if

$$f_X(y)f_Y(x) \leq f_X(x)f_Y(y)$$

for all $x < y$. The LR order can also be characterized by the following property:

$$X \leq_{LR} Y \Leftrightarrow (X|s < X \leq t) \leq_{ST} (Y|s < Y \leq t)$$

for all $s < t$ such that these conditional random variables exist (i.e., such that $F_X(s) < F_X(t)$ and $F_Y(s) < F_Y(t)$). This property can be used to give a general definition of the LR order. Hence the LR order can be interpreted as follows: $X \leq_{LR} Y$ if and only if when we know that a unit from X and another unit from Y have both failed in the interval $(s, t]$, the lifetime of the unit from X is ST-smaller than the lifetime of the unit from Y for all $s < t$. In particular, we obtain that the LR order implies both the HR and the RHR orders and, of course, the ST order. The LR order can also be characterized by the following property: $X \leq_{LR} Y$ holds iff $\eta_X \geq \eta_Y$, where $\eta_Z := -f'_Z/f_Z$ is known as the Glaser's eta function (see Glaser 1980).

The relationships between the preceding orders defined in terms of ST orderings of conditional random variables can be summarized as follows:

$$X \leq_{LR} Y \;\Rightarrow\; X \leq_{HR} Y$$
$$\Downarrow \qquad\qquad \Downarrow$$
$$X \leq_{RHR} Y \Rightarrow X \leq_{ST} Y$$

where the reverse implications are not necessarily true.

We can complete the diagram above by including the orders based on conditional expectations given below. For a random variable Z, we define the upper end-point u_Z of its support as $u_Z := \sup\{x : F_Z(x) < 1\}$. Analogously, the lower end-point l_Z of its support is $l_Z := \inf\{x : F_Z(x) > 0\}$. Then the mean residual lifetime order is defined as follows.

Definition 3.5 X is said to be smaller than Y in the **mean residual life order** (denoted by $X \leq_{MRL} Y$ of by $F_X \leq_{MRL} F_Y$) if $u_X \leq u_Y$ and $m_X(t) \leq m_Y(t)$ for all $t < u_X$ for which that expectations exist, where $m_X(t) = E(X - t|X > t)$ and $m_Y(t) = E(Y - t|Y > t)$ are the MRL functions of X and Y, respectively.

Analogously, we can define the following orders based on conditional expectations.

Definition 3.6 X is said to be smaller than Y in the **mean inactivity time order** (denoted by $X \leq_{MIT} Y$) if $l_X \leq l_Y$ and $\bar{m}_X(t) \geq \bar{m}_Y(t)$ for all $t > l_Y$ for which that expectations exist, where $\bar{m}_X(t) = E(t - X|X \leq t)$ and $\bar{m}_Y(t) = E(t - Y|Y \leq t)$ are the MIT functions of X and Y, respectively.

Definition 3.7 X is said to be smaller than Y in the **doubly truncated mean order** (denoted by $X \leq_{DTM} Y$) if $m_X(s,t) \leq m_Y(s,t)$ for all $s < t$ for which that expectations exist, where $m_X(s,t) = E(X|s < X \leq t)$ and $m_Y(s,t) = E(Y|s < Y \leq t)$ are the DTM functions of X and Y, respectively.

The definitions and relationships between the orders defined in this section can be summarized in the diagram given in Table 3.1 that was obtained by Navarro et al. (1997). The first and last columns can be used as definitions for general distributions. The implications from the second column to the third column are consequences of the characterization of the ST order given in (3.1). The other implications can be obtained taking limits to ∞ or to $-\infty$. The reverse implications are not necessarily true.

Another option two compare two independent random variables X and Y defined over the same probability space is the following.

Definition 3.8 If X and Y are two independent random variables defined over the same probability space, X is said to be smaller than Y in **stochastic precedence** (denoted by X SP Y) if $\Pr(X \leq Y) \geq 1/2$.

Table 3.1 Relationships between the main stochastic orders. We use the notation $Z_t = (Z - t | Z > t)$, $_tZ = (t - Z | Z \leq t)$ and $_sZ_t = (Z | s < Z \leq t)$

$$
\begin{array}{ccccccc}
\bar{F}_X \leq \bar{F}_Y & \Leftrightarrow & X \leq_{ST} Y & \Rightarrow & X \leq_M Y & \Leftrightarrow & E(X) \leq E(Y) \\
\Uparrow & & \Uparrow & & \Uparrow & & \Uparrow \\
_tX \geq_{ST} {}_tY & \Leftrightarrow & X \leq_{RHR} Y & \Rightarrow & X \leq_{MIT} Y & \Leftrightarrow & E(_tX) \geq E(_tY) \\
\Uparrow & & \Uparrow & & \Uparrow & & \Uparrow \\
_sX_t \leq_{ST} {}_sY_t & \Leftrightarrow & X \leq_{LR} Y & \Rightarrow & X \leq_{DTM} Y & \Leftrightarrow & E(_sX_t) \leq E(_sY_t) \\
\Downarrow & & \Downarrow & & \Downarrow & & \Downarrow \\
X_t \leq_{ST} Y_t & \Leftrightarrow & X \leq_{HR} Y & \Rightarrow & X \leq_{MRL} Y & \Leftrightarrow & E(X_t) \leq E(Y_t) \\
\Downarrow & & \Downarrow & & \Downarrow & & \Downarrow \\
\bar{F}_X \leq \bar{F}_Y & \Leftrightarrow & X \leq_{ST} Y & \Rightarrow & X \leq_M Y & \Leftrightarrow & E(X) \leq E(Y)
\end{array}
$$

It is an open problem to determine if stochastic precedence comparisons have the transitive property. They do not have it when X and Y are dependent. Hence we do not know if they define a proper order. For that reason we do not use the notation $X \leq_{SP} Y$. Moreover, note that if both X SP Y and Y SP X hold, then we do not know if X and Y have the same law. However, stochastic precedence is a reasonable way to compare the lifetimes of two independent units or systems. Moreover, Arcones et al. (2002) prove that if X and Y are two independent random variables defined over the same probability space and $X \leq_{ST} Y$ holds, then X SP Y. Hence stochastic precedence is a necessary condition for the ST order to hold. Stochastic precedence comparisons can be used as an alternative to the mean order when the ST order does not hold.

3.2 Systems with IID or EXC Components

First of all we prove that the k-out-of-n systems with IID components are LR-ordered (as expected).

Proposition 3.1 *If F is absolutely continuous, then*

$$X_{i:n} \leq_{LR} X_{j:m}$$

for all $i \leq j$ and $n - i \geq m - j$.

Proof From (2.12), we get

$$f_{i:n}(t) = i \binom{n}{i} f(t) F^{i-1}(t) \bar{F}^{n-i}(t)$$

and

$$f_{j:m}(t) = j \binom{m}{j} f(t) F^{j-1}(t) \bar{F}^{m-j}(t).$$

Hence

$$\frac{f_{j:m}(t)}{f_{i:n}(t)} = c\frac{F^{j-i}(t)}{\bar{F}^{n-i-m+j}(t)}$$

for a constant $c > 0$. As F is increasing and $\bar{F}$ is decreasing, this ratio is increaing in t under the stated assumptions and so the LR order holds.   □

As a consequence, in the IID case, we have that

$$X_{i:m} \leq_{LR} X_{i:n} \leq_{LR} X_{j:n}$$

for all $i, j, n, m \in \mathbb{Z}$ such that $1 \leq i \leq j \leq n \leq m$. Note that $X_{i:n}$ is LR increasing in i and LR decreasing in n. In particular, the k-out-of-n systems (order statistics) are LR ordered in the IID case, that is,

$$X_{1:n} \leq_{LR} \cdots \leq_{LR} X_{n:n}. \tag{3.4}$$

As the LR order is the strongest one, then

$$X_{1:n} \leq_{ORD} \cdots \leq_{ORD} X_{n:n} \tag{3.5}$$

for $ORD = HR, RHR, ST, MRL, MIT, DTM$. This property also hold if F is not absolutely continuous. Actually,

$$X_{1:n} \leq_{ST} \cdots \leq_{ST} X_{n:n} \tag{3.6}$$

holds in the general case since $X_{1:n} \leq \cdots \leq X_{n:n}$. In the general case we also have

$$X_{1:n} \leq_{ST} \cdots \leq_{ST} X_{1:1},$$

for the series systems,

$$X_{1:1} \leq_{ST} \cdots \leq_{ST} X_{n:n},$$

for the parallel systems and, in general, $X_{i:n} \leq_{ST} X_{j:m}$ whenever $i \leq j$ and $n - i \geq m - j$.

However, surprisingly, we will see that neither

$$X_{1:n} \leq_{HR} \cdots \leq_{HR} X_{n:n} \tag{3.7}$$

nor

$$X_{1:n} \leq_{MRL} \cdots \leq_{MRL} X_{n:n}. \tag{3.8}$$

hold in the general (or the EXC) case. This fact was first proved in Navarro and Shaked (2006).

Now we are ready to prove the first ordering results for systems with IID components based on Samaniego's signature representation. They were obtained in Kochar et al. (1999) and allows us to compare two systems just by comparing their respective signatures. Note that the signatures of order n can be considered as probability mass functions of discrete distributions over $\{1, \ldots, n\}$. Then they can be ordered by using the orders defined above.

Theorem 3.1 (Kochar et al. 1999) *Let T_1 and T_2 be the lifetimes of two coherent systems based on n IID components with a common continuous distribution function F. Let s_1 and s_2 be their respective signatures. Then the following properties hold:*

(i) *If* $s_1 \leq_{ST} s_2$, *then* $T_1 \leq_{ST} T_2$ *for all F;*

(ii) *If* $s_1 \leq_{HR} s_2$, *then* $T_1 \leq_{HR} T_2$ *for all F;*

(iii) *If* $s_1 \leq_{LR} s_2$, *then* $T_1 \leq_{LR} T_2$ *for all abs. cont. distribution functions F.*

The proof is obtained from Samaniego's representation (2.5), the ordering properties of the k-out-of-n systems in (3.5) for the IID case and the preservation ordering properties for mixtures of ordered distributions given in Theorems 1.A.6, 1.B.14 and 1.C.17 of Shaked and Shanthikumar (2007). Stochastic precedence comparisons were obtained in Theorem 5.6 of Samaniego (2007), p. 70.

These results can be extended to the EXC case by using the representations for coherent and semi-coherent systems obtained in the preceding chapter. These results were obtained in Navarro et al. (2008). Note that they also hold for systems with ID component lifetimes and a common DD survival copula due to Theorem 2.13.

Theorem 3.2 (Navarro et al. 2008) *Let T_1 and T_2 be the lifetimes of two semi-coherent (or coherent) systems with component lifetimes $(X_1, \ldots, X_n)$ having an exchangeable joint distribution function $\mathbf{F}$, and signatures of order n (signatures), $s_1^{(n)}$ and $s_2^{(n)}$, respectively. Then the following properties hold:*

(i) *If $s_1^{(n)} \leq_{ST} s_2^{(n)}$, then $T_1 \leq_{ST} T_2$ for all $\mathbf{F}$;*

(ii) *If $s_1^{(n)} \leq_{HR} s_2^{(n)}$, then $T_1 \leq_{HR} T_2$ for all $\mathbf{F}$ such that (3.7) holds;*

(iii) *If $s_1^{(n)} \leq_{HR} s_2^{(n)}$, then $T_1 \leq_{MRL} T_2$ for all $\mathbf{F}$ such that (3.8) holds;*

(iv) *If $s_1^{(n)} \leq_{LR} s_2^{(n)}$, then $T_1 \leq_{LR} T_2$ for all absolutely continuous or discrete joint distribution functions $\mathbf{F}$ such that (3.4) holds.*

As in the IID case, this theorem is an immediate consequence of the signature representation for the EXC case (2.27) and the mixture preservation properties obtained in Shaked and Shanthikumar (2007). However, in this case, we need to assume the respective ordering properties for the k-out-of-n systems (except in the case of the ST order where they are always true). Note that in (iii) we need the HR order for the signatures to get the MRL order for the system lifetimes when the k-out-of-n systems are MRL ordered. The MRL order for the signatures is not enough. Similar results holds for the MIT and RHR orders (see Navarro and Rubio 2011). Let us see an example.

Example 3.1 Let us consider the systems with lifetimes $T_1 = \min(X_1, \max(X_2, X_3))$ and $T_2 = \max(\min(X_1, X_2), \min(X_3, X_4))$. Note that they are of different orders (or that the first one is a semi-coherent system of order 4). So we need the signatures of order 4 to compare them. They are $s_1^{(4)} = (1/4, 5/12, 1/3, 0)$ and $s_2^{(4)} = s_2 = (0, 2/3, 1/3, 0)$, respectively. We also have to assume that (X_1, X_2, X_3, X_4) has an EXC joint distribution $\mathbf{F}$ (or that they are IID or just ID with a DD survival copula).

To check the ST order we need to compute the reliability functions $S_1^{(4)}$ and $S_2^{(4)}$ of the respective signatures. They are given in the following table:

$s_1^{(4)}$	1/4	5/12	1/3	0
$S_1^{(4)}$	1	3/4	1/3	0
$s_2^{(4)}$	0	2/3	1/3	0
$S_2^{(4)}$	1	1	1/3	0

As $S_1^{(4)} \le S_2^{(4)}$, then $s_1^{(4)} \le_{ST} s_2^{(4)}$ holds. Therefore, from Theorem 3.2, (i), $T_1 \le_{ST}$ T_2 holds for all EXC joint distributions F. This includes the IID$\sim F$ case for any univariate distribution function F. Note that these systems cannot be ordered by using Theorem 3.1.

Analogously, to check the HR order, we need to compute the ratio of the reliability functions $S_1^{(4)}$ and $S_2^{(4)}$ of the respective signatures. They are given in the following table:

$S_2^{(4)}$	1	1	1/3	0
$S_1^{(4)}$	1	3/4	1/3	0
$S_2^{(4)}/S_1^{(4)}$	1	4/3	1	–

Hence $s_1^{(4)}$ and $s_2^{(4)}$ are not HR ordered. Therefore, we do not know if T_1 and T_2 are HR ordered for all EXC joint distributions F such that (3.7) holds (or all F in the IID case). Note that Theorems 3.1 and 3.2 just include sufficient conditions for this ordering.

Finally, if we want to get the LR order, we need to compute the ratio of the respective signatures $s_1^{(4)}$ and $s_2^{(4)}$. It is given in the following table:

$s_2^{(4)}$	0	2/3	1/3	0
$s_1^{(4)}$	1/4	5/12	1/3	0
$s_2^{(4)}/s_1^{(4)}$	0	8/5	1	–

As expected, $s_1^{(4)}$ and $s_2^{(4)}$ are not LR ordered (since they are not HR ordered). So we do not know what happen with the system lifetimes in the LR order. Note again that Theorem 3.2 just includes sufficient conditions.

To illustrate these theoretical results we consider the IID case with a standard exponential distribution. The system reliability functions are plotted in Fig. 3.1, left. As expected they are ordered. This property holds for any distribution function F. Even more, it holds for any joint EXC distribution function F. The code in R to get this plot is the following:

```
R<-function(t) exp(-t)
s1<-c(1/4,5/12,1/3,0)
s2<-c(0,2/3,1/3,0)
R14<-function(t) (R(t))^4
R24<-function(t) 4*(R(t))^ 3-3*(R(t))^4
R34<-function(t) 6*(R(t))^2-8*(R(t))^3+3*(R(t))^ 4
```

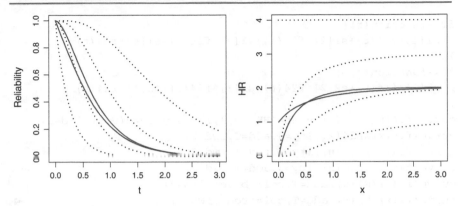

Fig. 3.1 Reliability functions (left) and hazard rate functions (right) for the systems T_1 (blue) and T_2 (red) in Example 3.1. The dotted lines correspond to the functions of the k-out-of-4 systems for $k = 1, 2, 3, 4$

```
R44<-function(t)  4*R(t)-6*(R(t))^2+4*(R(t))^ 3-1*(R(t))^4
R1<-function(t) {
  s1[1]*R14(t)+s1[2]*R24(t)+s1[3]*R34(t)+s1[4]*R44(t)
}
R2<-function(t) {
  s2[1]*R14(t)+s2[2]*R24(t)+s2[3]*R34(t)+s2[4]*R44(t)
}
curve(R14(x),xlab='t',ylab='Reliability',0,3,lty=3,lwd=2)
curve(R24(x),lty=3,add=T,lwd=2)
curve(R34(x),lty=3,add=T,lwd=2)
curve(R44(x),lty=3,add=T,lwd=2)
curve(R1(x),add=T,lwd=2)
curve(R2(x),add=T,col='red',lwd=2)
```

The system hazard rate functions are plotted in Fig. 3.1, right. In this case, they are not ordered. Thus, the second system is better when they are new but, from time $t = 0.5$ on (half a year if t is measured in years), the used systems with the first structure are a little bit better than that with the second. However, they have the same limiting behavior 2 when $t \to \infty$. Note that, in this example, the limiting behavior of the hazard rate function of the k-out-of-4 system is k for $k = 1, 2, 3, 4$ and that the common hazard rate of the components is $h(t) = 1$ for $t \geq 0$. The additional code to plot these hazard rate functions is the following:

```
f<-function(t)  exp(-t)
f14<-function(t)  f(t)*4*(R(t))^3
f24<-function(t)  f(t)*(12*(R(t))^2-12*(R(t))^3)
f34<-function(t)  f(t)*(12*R(t)-24*(R(t))^2+12*(R(t))^3)
f44<-function(t)  f(t)*(4-12*R(t)+12*(R(t))^2-4*(R(t))^3)
```

```
f1<-function(t){
  s1[1]*f14(t)+s1[2]*f24(t)+s1[3]*f34(t)+s1[4]*f44(t)
}
f2<-function(t) {
  s2[1]*f14(t)+s2[2]*f24(t)+s2[3]*f34(t)+s2[4]*f44(t)
}
curve(f14(x)/R14(x),ylab='HR',0,3,lty=3,ylim=c(0,4),lwd=2)
curve(f24(x)/R24(x),lty=3,add=T,lwd=2)
curve(f34(x)/R34(x),lty=3,add=T,lwd=2)
curve(f44(x)/R44(x),lty=3,add=T,lwd=2)
curve(f1(x)/R1(x),add=T,col='blue',lwd=2)
curve(f2(x)/R2(x),add=T,col='red',lwd=2)                    ◄
```

Proceeding as in the preceding example we can obtain all the ordering properties for all the coherent systems with 1-4 components given in Table 2.1. They were obtained in Navarro et al. (2008) and are given in Figs. 3.2, 3.3 and 3.4. The systems with repeated signatures are not included in the graphs (since they are equal in law to other systems in the graphs). Note that in the EXC case, we need some extra-conditions for the HR, MRL and LR orders. We do not need them in the IID case. Also note that the graph for the ST and LR orders are symmetric, that is, $T_i \leq_{ORD} T_j$ iff the respective dual systems satisfy $T_j^D \leq_{ORD} T_i^D$. This is not the case for the HR and MRL orders. For the hazard rate order, we have $T_i \leq_{HR} T_j$ iff $T_j^D \leq_{RHR} T_i^D$ under the respective properties for the order statistics in the EXC case. A similar property holds for MRL and MIT orders.

As we have seen in the preceding example, when the signature ordering does not hold, we do not know if the systems are ordered since the theorems just contain sufficient conditions. In Navarro and Rubio (2011) it is proved that the conditions given in (i), (ii) and (iv) of the preceding theorem are actually necessary and sufficient

Fig. 3.2 ST orderings for the systems in Table 2.1 and an EXC **F**. They also hold for the IID case

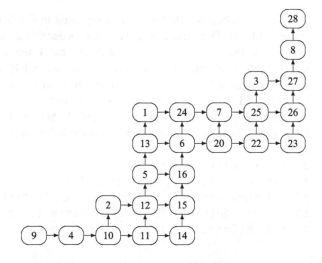

Fig. 3.3 HR (resp. MRL) orderings for the systems in Table 2.1 and an EXC *F* under (3.7) (resp. (3.8)). They also hold for the IID case (here both conditions hold and so it is better to get the HR order)

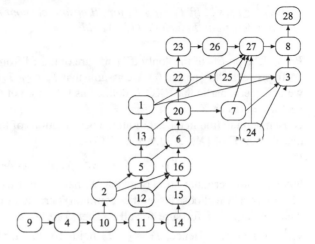

Fig. 3.4 LR orderings for the systems in Table 2.1 and and EXC *F* under (3.4). They also hold for the IID case for any absolute continuous distribution function *F*

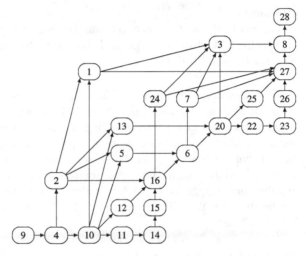

conditions to have the ST, HR and LR orderings, respectively, for any exchangeable distribution function *F* under these conditions for the order statistics. This result can be stated as follows. This is not true for the MRL order and, as we will see later, it is not true for the IID case.

Theorem 3.3 (Navarro and Rubio, 2011) *Let T_1 and T_2 be the lifetimes of two semi-coherent (or coherent) systems with component lifetimes $(X_1, \ldots, X_n)$ having an exchangeable joint distribution function $\mathbf{F}$, and signatures of order n (signatures), $\mathbf{s}_1^{(n)}$ and $\mathbf{s}_2^{(n)}$, respectively. Then the following properties hold:*

(i) $\mathbf{s}_1^{(n)} \leq_{ST} \mathbf{s}_2^{(n)}$ iff $T_1 \leq_{ST} T_2$ for all $\mathbf{F}$;

(ii) $\mathbf{s}_1^{(n)} \leq_{HR} \mathbf{s}_2^{(n)}$ iff $T_1 \leq_{HR} T_2$ for all $\mathbf{F}$ such that (3.7) holds;

(iii) $s_1^{(n)} \leq_{LR} s_2^{(n)}$ *iff* $T_1 \leq_{LR} T_2$ *for all absolutely continuous or discrete distribution functions* F *such that (3.4) holds.*

Proof The proofs of the "only if" parts are obtained from Theorem 3.2.

To prove the "if" part in (i), we assume that $T_1 \leq_{ST} T_2$ holds for all EXC F. Then we consider a particular EXC F defined as follows. Let $(X_1, \ldots, X_n)$ be a random vector defined as $X_i = \sigma(i)$ for $i = 1, \ldots, n$, where $\sigma : [n] \to [n]$ is a randomly chosen permutation in the set of all the permutations of the set $[n]$. Clearly, the joint distribution F of $(X_1, \ldots, X_n)$ is EXC since

$$(X_1, \ldots, X_n) =_{ST} (X_{\tau(1)}, \ldots, X_{\tau(n)})$$

holds for any permutation τ. Note that X_i has a uniform distribution on $[n]$ for all i. Moreover, as we choose a specific (common) permutation σ, the associated ordered data are $X_{i:n} = i$ for sure for all i. Therefore, $T_j = 1, \ldots, n$ with probabilities $s_j^{(n)}$ for $j = 1, 2$. Hence, $T_1 \leq_{ST} T_2$ holds for this EXC distribution F which is equivalent to $s_1^{(n)} \leq_{ST} s_2^{(n)}$.

The proofs of the "if" parts of (ii) and (iii) are analogous taking into account that the specific EXC distribution defined above trivially satisfies (3.7) and (3.4) (since $X_{i:n} = i$ for sure for all i). □

This theorem assures that Figs. 3.2, 3.3 and 3.4 contain all the ordering properties for the EXC case. Navarro and Rubio (2011) noticed a surprising property: Some systems that cannot be ordered by using signatures of order n, can be ordered with signatures of order m for some $m > n$. This fact seems to be against the preceding theorem but this is not the case. Let us see an example that proves that Fig. 3.4 does not contain all the ordering properties for the IID case. We will see in the next section that the same happen for Fig. 3.3, providing a procedure to detect all the orderings for the IID case. We will also prove that Fig. 3.2 does contain all the ST orderings for the IID case.

Example 3.2 Let us consider the systems 5 and 24 from Table 2.1 with lifetimes $T_5 = \min(X_1, \max(X_2, X_3))$ and $T_{24} = \max(X_1, \min(X_2, X_3, X_4))$. Their signatures of order 4 are $s_5^{(4)} = (1/4, 5/12, 1/3, 0)$ and $s_{24}^{(4)} = (0, 1/2, 1/4, 1/4)$. The respective reliability vectors are:

$s_5^{(4)}$	1/4	5/12	1/3	0
$S_5^{(4)}$	1	3/4	1/3	0
$s_{24}^{(4)}$	0	1/2	1/4	1/4
$S_{24}^{(4)}$	1	1	1/2	1/4
$S_{24}^{(4)}/S_5^{(4)}$	1	4/3	3/2	$+\infty$

As $S_{24}^{(4)}/S_5^{(4)}$ is increasing, $s_5^{(4)} \leq_{HR} s_{24}^{(4)}$ holds. So we can connect these systems in Fig. 3.3 and their respective lifetimes satisfy $T_5 \leq_{HR} T_{24}$ for all EXC F satisfying (3.7). In particular this ordering holds for the IID case and the ST order holds for any EXC F (note that $s_5^{(4)} \leq_{ST} s_{24}^{(4)}$ holds).

However, to check the LR order we compute the following table:

$s_{24}^{(4)}$	0	1/2	1/4	1/4
$s_5^{(4)}$	1/4	5/12	1/3	0
$s_{24}^{(4)}/s_5^{(4)}$	0	6/5	3/4	$+\infty$

Therefore, $s_5^{(4)}$ and $s_{24}^{(4)}$ are not LR-ordered. So these systems are not LR ordered for all EXC F and so they are not connected in Fig. 3.4.

However, if we compute the respective signatures of order 5 from (2.29), we get

$$s_5^{(5)} = \left(\frac{4}{5}\frac{1}{4}, \frac{1}{5}\frac{1}{4} + \frac{3}{5}\frac{5}{12}, \frac{2}{5}\frac{5}{12} + \frac{2}{5}\frac{1}{3}, \frac{3}{5}\frac{1}{3} + \frac{1}{5}0, \frac{4}{5}0\right) = \left(\frac{1}{5}, \frac{3}{10}, \frac{3}{10}, \frac{1}{5}, 0\right)$$

and

$$s_{24}^{(5)} = \left(\frac{4}{5}0, \frac{1}{5}0 + \frac{3}{5}\frac{1}{2}, \frac{2}{5}\frac{1}{2} + \frac{2}{5}\frac{1}{4}, \frac{3}{5}\frac{1}{4} + \frac{1}{5}\frac{1}{4}, \frac{4}{5}\frac{1}{4}\right) = \left(0, \frac{3}{10}, \frac{3}{10}, \frac{1}{5}, 0\right).$$

Hence,

$s_{24}^{(5)}$	0	3/10	3/10	1/5	1/5
$s_5^{(5)}$	1/5	3/10	3/10	1/5	0
$s_{24}^{(5)}/s_5^{(5)}$	0	1	1	1	$+\infty$

Therefore $s_5^{(5)} \leq_{LR} s_{24}^{(5)}$ holds and, from Theorem 3.2, (iv), $T_5 \leq_{LR} T_{24}$ for any EXC joint distribution F.

This property seems to contradict the property obtained with the signatures of order 4 taking into account that these properties are equivalent from Theorem 3.3, (iii). What is the explanation?

The answer is the following. Note that we have proved that $T_5 \leq_{LR} T_{24}$ for all EXC F of dimension 5. However, this is not true for all EXC F of dimension 4. In particular, this property fails for the distribution of dimension 4 constructed in the proof of Theorem 3.3 (since the systems' probability mass values $s_5^{(4)}$ and $s_{24}^{(4)}$ are not LR-ordered). This is due to the fact that this particular EXC distribution of dimension 4 cannot be extended (or included) in an exchangeable distribution of order 5. Note that we can affirm that $T_5 \leq_{LR} T_{24}$ holds for all EXC F of dimension 4 that can be extended (e.g. that are marginals) of EXC distributions of dimension 5. This is actually what happen in the IID case that can be extended to any dimension. So we can affirm that $T_5 \leq_{LR} T_{24}$ holds for the IID case and all distributions F (of dimension 1). Then note that we can connect these systems in the graph for the IID case. In the next section we will see how to complete the graphs for the IID case for all the orderings. ◀

3.3   Systems with ID Components

Recall that, from the preceding chapter, if the component lifetimes of a system are identically distributed with a common distribution F and a common reliability

$\bar{F} = 1 - F$, then the respective system's functions can be written as

$$F_T(t) = q(F(t)) \tag{3.9}$$

and

$$\bar{F}_T(t) = \bar{q}(\bar{F}(t)) \tag{3.10}$$

for all t, where q and $\bar{q}$ are two (univariate) distortion functions satisfying $\bar{q}(u) = 1 - q(1 - u)$ for all $u \in [0, 1]$. These distortion functions are increasing and continuous and depend on the system structure (minimal path or cut sets) and on the dependence between the component lifetimes (copula or survival copula).

Hence, we can apply to systems with ID components all the ordering properties obtained for distorted distributions in Navarro et al. (2013, 2014) and Navarro and Gomis (2016). They are stated in the following proposition. We say that a function g is **bathtub (upside-down bathtub)** shaped if there exist $t_1 \leq t_2$ such that $g(t)$ is decreasing (increasing) for $t \leq t_1$, constant for $t \in [t_1, t_2]$, and increasing (decreasing) for $t \geq t_2$. In many applications, the hazard rate functions of the components are bathtub shaped.

Proposition 3.2 *If T_i has the reliability function $\bar{q}_i(\bar{F}(t))$ and the distribution function $q_i(F(t))$ for $i = 1, 2$, then the following properties hold:*

(i) $T_1 \leq_{ST} T_2$ *for all F iff $\bar{q}_2 \geq \bar{q}_1$ (or $q_2 \leq q_1$) in $(0, 1)$;*
(ii) $T_1 \leq_{HR} T_2$ *for all F iff $\bar{q}_2/\bar{q}_1$ decreases in $(0, 1)$;*
(iii) $T_1 \leq_{RHR} T_2$ *for all F iff q_2/q_1 increases in $(0, 1)$;*
(iv) $T_1 \leq_{LR} T_2$ *for all absolutely continuous distribution functions F iff $\bar{q}_2'/\bar{q}_1'$ decreases (or q_2'/q_1' increases) in $(0, 1)$;*
(v) $T_1 \leq_{MRL} T_2$ *for all F such that $E(T_1) \leq E(T_2)$ if $\bar{q}_2/\bar{q}_1$ is bathtub in $(0, 1)$.*

Proof The proof (i) is immediate.

To prove (ii) we note that $T_1 \leq_{HR} T_2$ holds iff

$$\frac{\bar{q}_2(\bar{F}(t))}{\bar{q}_1(\bar{F}(t))}$$

is increasing in t. Clearly, this property holds when $\bar{q}_2/\bar{q}_1$ decreases in $(0, 1)$ since $\bar{F}$ is decreasing. Conversely, if $T_1 \leq_{HR} T_2$ holds for all F, then it holds for a continuous F (e.g. a standard exponential or a uniform distribution), and then $\bar{q}_2(u)/\bar{q}_1(u)$ is decreasing for $u \in (0, 1)$.

The proof of (iii) is similar to that of (ii).

To prove (iv) we recall that the respective PDF can be written as $f_i(t) = f(t)q_i'(F(t))$ for $i = 1, 2$, where $f = F'$ is the common baseline PDF. Hence, $T_1 \leq_{LR} T_2$ holds iff the ratio

$$\frac{f_2(t)}{f_1(t)} = \frac{\bar{q}_2'(\bar{F}(t))}{\bar{q}_1'(\bar{F}(t))}$$

is increasing in t. Clearly, this property holds when $\bar{q}_2'/\bar{q}_1'$ decreases in $(0, 1)$ since $\bar{F}$ is decreasing. Conversely, if $T_1 \leq_{LR} T_2$ holds for all F, then it holds for a continuous F and so $\bar{q}_2'/\bar{q}_1'$ decreases in $(0, 1)$. The proof for q_2'/q_1' is similar.

Finally, to prove (v), we note that if $\bar{q}_2/\bar{q}_1$ is bathtub in $(0, 1)$, then the ratio

$$\frac{\bar{F}_2(t)}{\bar{F}_1(t)} = \frac{\bar{q}_2(\bar{F}(t))}{\bar{q}_1(\bar{F}(t))}$$

is bathtub in t. Hence, from the results given in Belzunce et al. (2013), $T_1 \leq_{MRL} T_2$ holds for all F such that $E(T_1) \leq E(T_2)$.   □

Note that in all the orderings we have necessary and sufficient conditions except in (v) where we just have a sufficient condition and that there we need the additional condition $E(T_1) \leq E(T_2)$. Note that if $\bar{q}_2/\bar{q}_1$ is decreasing (or increasing), then we get the HR order from (ii) which is stronger than the MRL order. Moreover, we do not need the additional assumption $E(T_1) \leq E(T_2)$.

Clearly, these properties can be applied to compare systems with ID components having a common distribution function F by using the distortion representations obtained in Sect. 2.4. The result for the ST order can be stated as follows.

Proposition 3.3 *Let T_1 and T_2 be the lifetimes of two semi-coherent (or coherent) systems with ID component lifetimes having an common distribution function F, and distortion functions q_1 and q_2, respectively. Then the following properties are equivalent:*

(i) $\bar{q}_1 \leq \bar{q}_2$ (or $q_1 \geq q_2$) in $(0, 1)$;
(ii) $T_1 \leq_{ST} T_2$ for all F;
(iii) $T_1 \leq_{ST} T_2$ for a continuous F.

Proof From Proposition 3.2, (i), we have that (i) implies (ii).

Clearly, (ii) implies (iii).

Finally, if (iii) holds, then $T_1 \leq_{ST} T_2$ for a continuous F, that is, $\bar{F}_{T_1} \leq \bar{F}_{T_2}$. Hence, if $0 < u < 1$, then there exists t such that $\bar{F}(t) = u$ (since F is continuous). Therefore

$$\bar{q}_1(u) = \bar{q}_1(\bar{F}(t)) = \bar{F}_{T_1}(t) \leq \bar{F}_{T_2}(t) = \bar{q}_2(\bar{F}(t)) = \bar{q}_2(u)$$

for all $u \in (0, 1)$.   □

For the HR order we have the following result.

Proposition 3.4 *Let T_1 and T_2 be the lifetimes of two semi-coherent (or coherent) systems with ID component lifetimes having a common distribution function F, and distortion functions q_1 and q_2, respectively. Then the following properties are equivalent:*

(i) $\bar{q}_2/\bar{q}_1$ is decreasing in $(0, 1)$;

(ii) $T_1 \leq_{HR} T_2$ for all F;
(iii) $T_1 \leq_{HR} T_2$ for a continuous F.

The proof is similar to that of the ST order. Note that in both cases, the systems may have different orders (i.e. numbers of components), different structures and different dependency relationships (copulas). The only requirement is that they have a common distribution function F. Also note that then we get distribution-free ordering results, that is, comparisons for any F. Similar results can be stated for the RHR and LR orders from Proposition 3.2. In the last case we need to assume that the respective distortion functions are differentiable. However, the result for the MRL ordering is different. It can be stated as follows.

Proposition 3.5 *Let T_1 and T_2 be the lifetimes of two semi-coherent (or coherent) systems with ID component lifetimes having an common distribution function F, and distortion functions q_1 and q_2, respectively. If $\bar{q}_2/\bar{q}_1$ is bathtub in $(0, 1)$, then $T_1 \leq_{MRL} T_2$ for all F such that $E(T_1) \leq E(T_2)$.*

The converse property does not hold (for strict bathtub shaped functions, that is, with both strict decreasing and strict increasing pieces). A counterexample can be seen in Navarro and Gomis (2016). Let us see how to apply the preceding results to systems with dependent ID components.

Example 3.3 Let us consider a series system and a parallel system with ID components having a common reliability $\bar{F}$ and a survival copula $\widehat{C}$. The reliability function of the series system $X_{1:2}$ can be written as

$$\bar{F}_{1:2}(t) = \Pr(X_{1:2} > t) = \Pr(X_1 > t, X_2 > t) = \widehat{C}(\bar{F}(t), \bar{F}(t)) = \bar{q}_{1:2}(\bar{F}(t)),$$

where $q_{1:2}(u) = \widehat{C}(u, u)$ is the diagonal section of the copula $\widehat{C}$.

Analogously, the reliability function of the parallel system $X_{2:2}$ is

$$\begin{aligned}
\bar{F}_{2:2}(t) &= \Pr(\max(X_1, X_2) > t) \\
&= \Pr(X_1 > t) + \Pr(X_2 > t) - \Pr(X_1 > t, X_2 > t) \\
&= 2\bar{F}(t) - \widehat{C}(\bar{F}(t), \bar{F}(t)) \\
&= \bar{q}_{2:2}(\bar{F}(t)),
\end{aligned}$$

where $q_{2:2}(u) = 2u - \widehat{C}(u, u)$ for $u \in [0, 1]$.

Note that, in this case (and in the general case), we know that

$$X_{1:2} \leq_{ST} X_i \leq_{ST} X_{2:2}$$

holds for $i = 1, 2$, for all F and for all $\widehat{C}$.

From Proposition 3.4, $X_{1:2} \leq_{HR} X_i$ holds for all $\bar{F}$ iff the ratio

$$\frac{\bar{q}_{1:2}(u)}{\bar{q}_i(u)} = \frac{\widehat{C}(u, u)}{u}$$

is increasing in $(0, 1)$. In a similar way, $X_i \leq_{HR} X_{2:2}$ holds for all F iff

$$\frac{\bar{q}_{2:2}(u)}{\bar{q}_i(u)} = \frac{2u - \widehat{C}(u, u)}{u}$$

decreases in $(0, 1)$, that is, iff $\widehat{C}(u, u)/u$ is increasing in $(0, 1)$. Analogously, it can also be proved that $X_{1:2} \leq_{HR} X_{2:2}$ holds for all F iff the same condition holds (i.e. $\widehat{C}(u, u)/u$ is increasing in $(0, 1)$). Therefore, in the ID case, these orderings are equivalent and they will just depend on the copula $\widehat{C}$ (they are distribution-free with respect to F).

Of course, if the components are IID, that is, $\widehat{C}(u, v) = uv$ for $u, v \in [0, 1]$, then $\widehat{C}(u, u)/u = u$, which is increasing, and so

$$X_{1:2} \leq_{HR} X_i \leq_{HR} X_{2:2} \tag{3.11}$$

holds for $i = 1, 2$ and for all F. This is a well known property already obtained in the preceding section (by using the LR order).

Analogously, if we consider the following Clayton–Oakes copula

$$\widehat{C}(u, v) = \frac{uv}{u + v - uv}, \quad u, v \in [0, 1], \tag{3.12}$$

which induces a positive dependence between the components, we get

$$\frac{\widehat{C}(u, u)}{u} = \frac{u^2}{2u^2 - u^3} = \frac{1}{2 - u}$$

which is increasing in $(0, 1)$. So (3.11) holds for all F and this copula. Of course, the same MRL orderings also hold for any F. However, there exist copulas such that this condition does not hold (see Example 4.1 in Navarro et al. 2018).

Let us study now the LR orderings. Thus, $X_{1:2} \leq_{LR} X_i$ holds for all F iff $\bar{q}'_{1:2}(u)/\bar{q}'_i(u) = \bar{q}'_{1:2}(u)$ is increasing in $(0, 1)$, that is, when $\bar{q}_{1:2}(u)$ is convex in $(0, 1)$. This is also the condition for the other LR orderings. In the IID case $\bar{q}_{1:2}(u) = u^2$ is convex in $(0, 1)$. Thus we can prove again that

$$X_{1:2} \leq_{LR} X_i \leq_{LR} X_{2:2} \tag{3.13}$$

holds for any F in the IID case. For the copula (3.12), we note that

$$\bar{q}_{1:2}(u) = \widehat{C}(u, u) = \frac{u}{2 - u}$$

is convex in $(0, 1)$ and so (3.13) holds for any F.

To illustrate these theoretical results we consider a standard exponential distribution F, and then we plot in Fig. 3.5 the reliability functions (left) and the hazard rate functions (right) of these systems for the IID case (dashed lines) and the copula in (3.12) (continuous lines). The R-code to get these plots is the following:

```
# Reliability functions
#IID case:
R<-function(t) exp(-t)
qIID<-function(u) u^2
G12<-function(t) qIID(R(t))
```

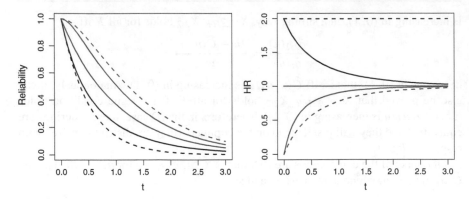

Fig. 3.5 Reliability (left) and hazard rate functions (right) for the series system $X_{1:2}$ (black), the components X_i (red) and the parallel system $X_{2:2}$ (blue) in Example 3.3 for the case of IID components (dashed lines) and dependent ID components (continuous lines) with the survival copula in (3.12)

```
G22<-function(t) 2*R(t)-G12(t)
curve(G12(x),xlab='t',ylab='Reliability',0,3,lty=2,lwd=2)
curve(G22(x),add=T,col='blue',lty=2,lwd=2)
curve(R(x),add=T,col='red',lwd=2)
#ID-C case:
C<-function(u,v) u*v/(u+v-u*v)
q<-function(u) C(u,u)
R12<-function(t) q(R(t))
R22<-function(t) 2*R(t)-R12(t)
curve(R12(x),xlab='t',add=T,lwd=2)
curve(R22(x),add=T,col='blue',lwd=2)
curve(R(x),add=T,col='red',lwd=2)


# Hazard rate functions
#IID case:
f<-function(t) exp(-t)
qpIID<-function(u) 2*u
g12<-function(t) f(t)*qpIID(R(t))
g22<-function(t) 2*f(t)-g12(t)
curve(g12(x)/G12(x),ylab='HR',0,3,ylim=c(0,2),lty=2,lwd=2)
curve(g22(x)/G22(x),add=T,col='blue',lty=2,lwd=2)
curve(f(x)/R(x),add=T,col='red',lwd=2)
#ID-C case:
qp<-function(u) 2/(2-u)^2
f12<-function(t) f(t)*qp(R(t))
f22<-function(t) 2*f(t)-f12(t)
```

```
curve(f12(x)/R12(x),add=T,lwd=2)
curve(f22(x)/R22(x),add=T,col='blue',lwd=2)
curve(f(x)/R(x),add=T,col='red',lwd=2)
```

Analogously, we can compare the systems obtained in the IID case with that obtained with the copula $\widehat{C}$. For example, $X_{1:2}^{IID} \leq_{HR} X_{1:2}^{\widehat{C}}$ holds for all F since

$$\frac{\bar{q}_{1:2}^{\widehat{C}}(u)}{\bar{q}_{1:2}^{IID}(u)} = \frac{u/(2-u)}{u^2} = \frac{1}{2u - u^2}$$

is decreasing in $(0, 1)$. Analogously, we get that $X_{2:2}^{IID} \geq_{HR} X_{2:2}^{\widehat{C}}$ holds for all F since

$$\frac{\bar{q}_{2:2}^{\widehat{C}}(u)}{\bar{q}_{2:2}^{IID}(u)} = \frac{2u - u/(2-u)}{2u - u^2} = \frac{3 - 2u}{(2-u)^2}$$

is increasing in $(0, 1)$. Note that the series system improves with the positive dependency but that the parallel system get worse (see Fig. 3.5). ◀

As we have seen in the preceding example, the (distribution-free) ordering properties between two systems with ID components will just depend on the copula, that is, the dependence structure. So they can be related to well known positive/negative dependence properties. These relationships where studied in Navarro et al. (2018) and Navarro et al. (2021). For instance, the results obtained in the preceding example for series and parallel systems with two ID components can be stated as follows.

Proposition 3.6 *Let X_1 and X_2 be the lifetimes of two components having a common distribution function F and copula and survival copula C and $\widehat{C}$, respectively. Then the following properties are equivalent:*

  (i) $X_{1:2} \leq_{HR} X_1$ *for all F;*
 (ii) $X_1 \leq_{HR} X_{2:2}$ *for all F;*
(iii) $X_{1:2} \leq_{HR} X_{2:2}$ *for all F;*
 (iv) $\widehat{C}(u, u)/u$ *is increasing in $(0, 1)$;*
  (v) $(1 - C(u, u))/(1 - u)$ *is increasing in $(0, 1)$.*

Note that to prove (iv) (or (v)) we just need one of these orderings for a continuous distribution function F. Also note that in the ID case, as $\bar{F}_{2:2} = 2\bar{F} - \bar{F}_{1:2}$, then

$$\bar{F}(t) = \frac{1}{2}\bar{F}_{1:2}(t) + \frac{1}{2}\bar{F}_{2:2}(t)$$

for all t, that is, the common components' distribution is a uniform mixture of the distributions of the series and the parallel system. So the HR function of the components will be always between that of series and parallel systems (for any copula). This fact explains why the orderings stated in the preceding proposition are equivalent. For the LR order, we have the following result. The conditions for the RHR order can be seen in Theorem 4.2 of Navarro et al. (2018).

Proposition 3.7 *Let X_1 and X_2 be the lifetimes of two components having a common absolutely continuous distribution function F and copula and survival copula C and $\widehat{C}$, respectively. Then the following properties are equivalent:*

(i) $X_{1:2} \leq_{LR} X_1$ *for all F;*
(ii) $X_1 \leq_{LR} X_{2:2}$ *for all F;*
(iii) $X_{1:2} \leq_{LR} X_{2:2}$ *for all F;*
(iv) $\widehat{C}(u, u)$ *is convex in* $(0, 1)$.
(v) $C(u, u)$ *is convex in* $(0, 1)$.

Analogously, the condition for the comparisons of the IID case with the DID case are the following. They were obtained in Proposition 17 of Navarro et al. (2021).

Proposition 3.8 *Let X_1 and X_2 be the lifetimes of two components having a common distribution function F and survival copula $\widehat{C}$. Let $\delta_{\widehat{C}}(u) = \widehat{C}(u, u)$ for $u \in [0, 1]$. Let Y_1 and Y_2 be two IID lifetimes with distribution F.*

(i) $Y_{1:2} \leq_{ST} X_{1:2}$ $(\geq_{ST})$ *for all F iff* $u^2 \leq \delta_{\widehat{C}}(u)$ $(\geq)$ *for all $u \in (0, 1)$;*
(ii) $Y_{1:2} \leq_{HR} X_{1:2}$ $(\geq_{HR})$ *for all F iff* $\delta_{\widehat{C}}(u)/u^2$ *is decreasing (increasing) in* $(0, 1)$;
(iii) $Y_{1:2} \leq_{LR} X_{1:2}$ $(\geq_{LR})$ *for all abs. cont. F iff* $\delta'_{\widehat{C}}(u)/u$ *is decreasing (increasing) in* $(0, 1)$;
(iv) $Y_{2:2} \geq_{ST} X_{2:2}$ $(\leq_{ST})$ *for all F iff* $u^2 \leq \delta_{\widehat{C}}(u)$ $(\geq)$ *for all $u \in (0, 1)$;*
(v) $Y_{2:2} \geq_{HR} X_{2:2}$ $(\leq_{HR})$ *for all F iff* $(2u - \delta_{\widehat{C}}(u))/(2u - u^2)$ *is increasing (decreasing) in* $(0, 1)$;
(vi) $Y_{2:2} \geq_{LR} X_{2:2}$ $(\leq_{LR})$ *for all abs. cont. F iff* $(2 - \delta'_{\widehat{C}}(u))/(1 - u)$ *is increasing (decreasing) in* $(0, 1)$.

Note that $Y_{1:2} \leq_{ST} X_{1:2}$ $(\geq_{ST})$ holds iff $Y_{2:2} \geq_{ST} X_{2:2}$ $(\leq_{ST})$. However, the other orderings are not equivalent. A random vector (X_1, X_2) is Positive (Negative) Quadrant Dependent, shortly written as PQD (NQD), if $F(x, y) \geq F_1(x)F_2(y)$ $(\leq)$ for all x, y, (see, e.g., Joe 1997). If F_1, F_2 are continuous, these (dependence) properties only depend on the copula.

Proposition 3.9 *Let X_1 and X_2 be the two random variables having distribution functions F_1 and F_2 and copula and survival copula C and $\widehat{C}$, respectively. Then the following properties are equivalent:*

(i) (X_1, X_2) *is PQD (NQD) for all F_1, F_2;*
(ii) (X_1, X_2) *is PQD (NQD) for two continuous distributions F_1, F_2;*
(iii) $C(u, v) \geq uv$ $(\leq)$ *for all $u, v \in [0, 1]$;*
(iv) $\widehat{C}(u, v) \geq uv$ $(\leq)$ *for all $u, v \in [0, 1]$.*

Note that, in the ID case, $Y_{1:2} \leq_{ST} X_{1:2}$ $(\geq_{ST})$ and $Y_{2:2} \geq_{ST} X_{2:2}$ $(\leq_{ST})$ hold for all F when (X_1, X_2) is PQD (NQD). Thus the series system is better under a

Fig. 3.6 All the HR
orderings for the systems in
Table 2.1 and IID
components. The red arrow
is an ordering that cannot be
obtained by using signatures
of order 4

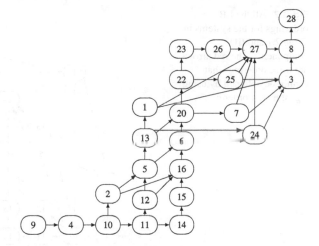

positive dependence but the opposite holds for the parallel system (as we have seen in the preceding example). These orderings are reverted for the NQD condition. The other conditions can also be related with dependence properties (see next section). These are expectable properties (series systems improve under positive dependence since both component lifetimes are similar while parallel systems does so when they are different).

Note that the necessary and sufficient conditions obtained above can also be used to obtain all the distribution-free comparisons of coherent (or semi-coherent) systems with IID components. All the orderings for systems with 1-4 components (given in Table 2.1) were obtained in Navarro (2016). In some cases, these results improve the results obtained by using signatures (see the preceding section). For example, for the HR order we obtain the relationships given in Fig. 3.6. Note that we have a new ordering $(13 \rightarrow 24)$ that cannot be obtained from signatures of order 4 (see Fig. 3.3). Analogously, for the LR order, we obtain the relationships given in Fig. 3.7. Note that we have three new orderings $(13 \rightarrow 24, 5 \rightarrow 24$ and $13 \rightarrow 7)$ that cannot be obtained from signatures of order 4 (see Fig. 3.4). For the ST order we obtain the same orderings given in the preceding section (see Fig. 3.2). However, for $n = 5$ and $n = 6$, there exist systems that can be ST-ordered with distortion functions but that cannot be ordered with signatures (see Rychlik et al. 2018). Moreover, note that the results based on distortions can also be used to check the ordering conditions for k-out-of-n systems needed in the ordering results based on signatures for the EXC case. For example, for $n = 3$, we can check if $X_{1:3} \leq_{HR} X_{2:3} \leq_{HR} X_{3:3}$ holds for a given copula C.

In other situations we may want to study if, for a fixed system (structure) and a fixed dependence (copula), an order is preserved. Thus, if the components $X_1, \ldots, X_n$ are ID$\sim F$ and $Y_1, \ldots, Y_n$ are ID$\sim G$, they share the same copula C and $F \leq_{ORD} G$ holds, we want to study if $T_1 \leq_{ORD} T_2$ holds (or holds under some conditions) for a given order ORD, where $T_1 = \phi(X_1, \ldots, X_n)$ and $T_2 = \phi(Y_1, \ldots, Y_n)$ are the lifetimes of two systems having the same structure.

Fig. 3.7 All the LR orderings for the systems in Table 2.1 and IID components. The red arrows are three orderings that cannot be obtained by using signatures of order 4

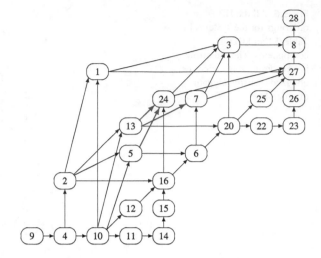

To this end we can use the following ordering results for distorted distributions extracted from Navarro et al. (2013). The similar results for the non-ID case were obtained in Navarro et al. (2016).

Proposition 3.10 *Let X and Y be the two random variables having absolutely continuous distribution functions F_X and F_Y. Let T and S be two random variables having distribution functions $q(F_X)$ and $q(F_Y)$ for a distortion function q. Let $\bar{q}$ be the dual distortion function and let $\alpha(u) = u\bar{q}'(u)/\bar{q}(u)$, $\tilde{\alpha}(u) = uq'(u)/q(u)$, and $\beta(u) = u\bar{q}''(u)/\bar{q}(u)$.*

(i) If $X \leq_{ST} Y$, then $T \leq_{ST} S$;
(ii) If $X \leq_{HR} Y$ and α is decreasing in $(0, 1)$, then $T \leq_{HR} S$;
(iii) If $X \leq_{RHR} Y$ and $\tilde{\alpha}$ is increasing in $(0, 1)$, then $T \leq_{RHR} S$;
(iv) If $X \leq_{LR} Y$ and β is non-negative and decreasing in $(0, 1)$, then $T \leq_{LR} S$.

Proof The proof of (i) is immediate (since q and $\bar{q}$ are increasing functions).

To prove (ii), we assume $X \leq_{HR} Y$, that is, $h_X \geq h_Y$ holds for the respective hazard rate functions. Hence, $X \leq_{ST} Y$ also holds, that is, $\bar{F}_X \leq \bar{F}_Y$. Then we use (2.34) and that α is decreasing and non-negative to get

$$h_T(t) = \alpha(\bar{F}_X(t))h_X(t) \geq \alpha(\bar{F}_Y(t))h_Y(t) = h_S(t)$$

for all t, for the respective hazard rate functions of T and S. Then $T \leq_{HR} S$ holds.

The proof of (iii) is similar to the preceding one from (2.35).

Finally, to prove (iv), we note that $X \leq_{LR} Y$ implies $\eta_X \geq \eta_Y$ for the respective Glaser's eta functions defined in the first section of this chapter. Moreover, $X \leq_{LR} Y$ implies $X \leq_{HR} Y$ (i.e. $h_X \geq h_Y$) and $X \leq_{ST} Y$ (i.e. $\bar{F}_X \leq \bar{F}_Y$). Then we use (2.33) and that β is decreasing and non-negative to get

$$\eta_T(t) = \eta_X(t) + \beta(\bar{F}(t))h_X(t) \geq \eta_Y(t) + \beta(\bar{F}_Y(t))h_Y(t) = h_S(t)$$

for all t, for the respective Glaser's eta functions of T and S. Then $T \leq_{LR} S$ holds. $\square$

Note that the ST-order is always preserved. However, we need some conditions for the preservations of the other orders. An alternative condition for the preservation of the HR is that the function $\bar{q}(uv)/\bar{q}(u)$ is increasing in $(0, 1)^2$. It can be proved that the HR order is preserved in k-out-of-n systems with IID components (i.e. α is decreasing for that systems). However, this is not the case for other coherent systems. Let us see an example.

Example 3.4 Let us consider the two systems with lifetimes T_1 and T_2, with a common structure $\phi(x_1, x_2, x_3) = \max(x_1, \min(x_2, x_3))$ and with IID components having distribution functions F and G, respectively. Then the common dual distortion function for these systems is

$$\bar{q}(u) = u + u^2 - u^3.$$

Hence,

$$\alpha(u) = u \frac{1 + 2u - 3u^2}{u + u^2 - u^3} = \frac{1 + 2u - 3u^2}{1 + u - u^2}.$$

By plotting α in $[0, 1]$, we see that it is non monotone (it first increases and then decreases). Therefore we do not know if the HR order is preserved. For example let us consider IID components having the reliability function

$$\bar{F}(t; a) = 1 - (1 - e^{-t})^a, \ t \geq 0 \tag{3.14}$$

for $a = 2, 5$. Then we plot in Fig. 3.8, left, the reliability functions of the components $\bar{F}(t) = \bar{F}(t; 2)$ (black dashed lines) and $\bar{G}(t) = \bar{F}(t; 5)$ (red dashed lines) and that of the respective systems (black and red continuous lines). As we can see, the components are ST ordered and this order is preserved in the systems (i.e. the system with the most reliable component, is more reliable than the other). In Fig. 3.8, right, we plot the hazard rate functions of the components (dashed lines) and the systems (continuous lines). As we can see, $F \leq_{HR} G$ holds. However, the hazard rate functions of the systems are not ordered. The code in R to get these plots is the following:

```
# Reliability functions:
R1<-function(t) 1-(1-exp(-t))^2
R2<-function(t) 1-(1-exp(-t))^5
q<-function(u) u+u^2-u^3
RT1<-function(t) q(R1(t))
RT2<-function(t) q(R2(t))
curve(RT1(x),xlab='t',ylab='Reliability',0,7,lwd=2)
curve(RT2(x),add=T,col='red',lwd=2)
curve(R1(x),add=T,lty=2,lwd=2)
curve(R2(x),add=T,col='red',lty=2,lwd=2)
```

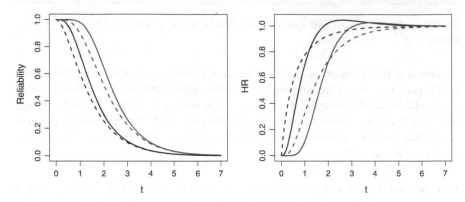

Fig. 3.8 Reliability functions (left) and hazard rate functions (right) for the components (dashed lines) and the systems (continuous lines) in Example 3.4 for the case of IID components with the reliability function in (3.14) and $a = 2$ (black) and $a = 5$ (red)

```
# Hazard rate functions:
f1<-function(t)  2*exp(-t)*(1-exp(-t))
f2<-function(t)  5*exp(-t)*(1-exp(-t))^4
qp<-function(u)  1+2*u-3*u^2
fT1<-function(t)  f1(t)* qp(R1(t))
fT2<-function(t)  f2(t)* qp(R2(t))
curve(fT1(x)/RT1(x),xlab='t',ylab='HR',0,7,lwd=2)
curve(fT2(x)/RT2(x),add=T,col='red',lwd=2)
curve(f1(x)/R1(x),add=T,lty=2,lwd=2)
curve(f2(x)/R2(x),add=T,col='red',lty=2,lwd=2)                ◀
```

3.4 Systems with Non-ID Components

First, we recall that, from the representation results obtained in the preceding chapter, the system distribution function can be written (in the general case) as

$$F_T(t) = Q(F_1(t), \ldots, F_n(t)),$$

and its reliability function as

$$\bar{F}_T(t) = \bar{Q}(\bar{F}_1(t), \ldots, \bar{F}_n(t)),$$

that is, they are generalized distorted distributions from the distributions of the component lifetimes. The explicit expression for the distortion functions Q and $\bar{Q}$ can be obtained from the minimal path (or cut) sets representation and the survival copula $\widehat{C}$ (or the copula C). So they only depend on the structure function and the dependence between the components (i.e. they do not depend on $F_1, \ldots, F_n$).

If the components are independent (IND), then the function $\bar{Q}$ is a multinomial and it is known as the **reliability function of the structure** (see Barlow and Proschan 1975, p. 21). Actually, this multinomial is the one obtained in the pivotal decomposition (1.3) or in representation based on the Möbius transform (1.10) when these Boolean functions are extended to real numbers. In this case Q is also a multinomial.

In both cases we can use the following ordering results for generalized distorted distributions obtained in Navarro et al. (2016) (arbitrary components) and in Navarro and del Águila (2017) (ordered components). Note that we have necessary and sufficient conditions for the ST, HR and RHR orders. In Navarro et al. (2016) there are sufficient conditions for the LR order.

Theorem 3.4 *If T_i has the distribution function $Q_i(F_1, \ldots, F_n)$ and the reliability function $\bar{Q}_i(\bar{F}_1, \ldots, \bar{F}_n)$, for $i = 1, 2$, then the following properties hold:*

(i) *$T_1 \leq_{ST} T_2$ for all $F_1, \ldots, F_n$ iff $\bar{Q}_1 \leq \bar{Q}_2$ (or $Q_1 \geq Q_2$) in $(0, 1)^n$;*
(ii) *$T_1 \leq_{HR} T_2$ for all $F_1, \ldots, F_n$ iff $\bar{Q}_2/\bar{Q}_1$ is decreasing in $(0, 1)^n$;*
(iii) *$T_1 \leq_{RHR} T_2$ for all $F_1, \ldots, F_n$ iff Q_2/Q_1 is increasing in $(0, 1)^n$.*

Proof The proof of (i) is immediate.

To prove (ii) we note that $T_1 \leq_{HR} T_2$ holds iff

$$\frac{\bar{F}_{T_2}(t)}{\bar{F}_{T_1}(t)} = \frac{\bar{Q}_2(\bar{F}_1(t), \ldots, \bar{F}_n(t))}{\bar{Q}_1(\bar{F}_1(t), \ldots, \bar{F}_n(t))} \tag{3.15}$$

is increasing in t.

If this ordering holds for all $F_1, \ldots, F_n$ and we want to prove that $\bar{Q}_2/\bar{Q}_1$ is decreasing in u_1 for fixed $u_2, \ldots, u_n \in (0, 1)$, we choose distribution functions such that $\bar{F}_i(t) = u_i$ for $t \in (1, 2)$ and $i = 2, \ldots, n$ and $\bar{F}_1(t) = 1$ for $t \leq 1$, $\bar{F}_1(t) = 2 - t$ for $t \in (1, 2)$, and $\bar{F}_1(t) = 0$ for $t \geq 2$. Then, from (3.15), we have that $\bar{Q}_2(u_1, \ldots, u_n)/\bar{Q}_1(u_1, \ldots, u_n)$ is decreasing in u_1. We can prove that it is decreasing in the other variables in a similar way.

Conversely, if we assume that $\bar{Q}_2/\bar{Q}_1$ is decreasing in all its variables in $(0, 1)^n$, as $\bar{F}_1, \ldots, \bar{F}_n$ are decreasing, from (3.15), $\bar{F}_{T_2}(t)/\bar{F}_{T_1}(t)$ is increasing in t.

The proof of (iii) is similar to the proof of (ii). $\square$

Theorem 3.5 *If T_i has the distribution function $Q_i(F_1, \ldots, F_n)$ and the reliability function $\bar{Q}_i(\bar{F}_1, \ldots, \bar{F}_n)$, for $i = 1, 2$, then the following properties hold:*

(i) *$T_1 \leq_{ST} T_2$ for all $F_1, \ldots, F_n$ such that $F_1 \geq_{ST} \cdots \geq_{ST} F_n$ iff $\bar{Q}_1 \leq \bar{Q}_2$ in $D = \{(u_1, \ldots, u_n) \in [0, 1]^n : u_1 \geq \cdots \geq u_n\}$;*
(ii) *$T_1 \leq_{HR} T_2$ for all $F_1, \ldots, F_n$ such that $F_1 \geq_{HR} \cdots \geq_{HR} F_n$ iff the function*

$$\bar{H}(v_1, \ldots, v_n) = \frac{\bar{Q}_2(v_1, v_1 v_2, \ldots, v_1 \ldots v_n)}{\bar{Q}_1(v_1, v_1 v_2, \ldots, v_1 \ldots v_n)} \tag{3.16}$$

is decreasing in $(0, 1)^n$;

(iii) $T_1 \leq_{RHR} T_2$ *for all* $F_1, \ldots, F_n$ *such that* $F_1 \leq_{RHR} \cdots \leq_{RHR} F_n$ *iff the function*

$$H(v_1, \ldots, v_n) = \frac{Q_2(v_1, v_1 v_2, \ldots, v_1 \ldots v_n)}{Q_1(v_1, v_1 v_2, \ldots, v_1 \ldots v_n)} \tag{3.17}$$

is increasing in $(0, 1)^n$.

Proof The proof of (i) is immediate since $F_1 \geq_{ST} \cdots \geq_{ST} F_n$ implies $\bar{F}_1 \geq \cdots \geq \bar{F}_n$.

To prove (ii) we recall that $T_1 \leq_{HR} T_2$ holds iff the ratio in (3.15) is increasing in t.

If we want to prove that this ordering holds for all $F_1 \geq_{HR} \cdots \geq_{HR} F_n$ when $\bar{H}$ is decreasing, we note $r_i = \bar{F}_i / \bar{F}_{i-1}$ is decreasing for $i = 2, \ldots, n$. Therefore, $r_i \in [0, 1]$ (since $r_i(0) = 1$). Moreover, $\bar{F}_1 \in [0, 1]$ and it is also decreasing. Hence,

$$\bar{H}(\bar{F}_1(t), r_2(t), \ldots, r_n(t)) = \frac{\bar{Q}_2(\bar{F}_1(t), \bar{F}_2(t), \ldots, \bar{F}_n(t))}{\bar{Q}_1(\bar{F}_1(t), \bar{F}_2(t), \ldots, \bar{F}_n(t))}$$

is increasing in t and so $T_1 \leq_{HR} T_2$ holds.

Conversely, let us assume that $T_1 \leq_{HR} T_2$ holds for all $F_1 \geq_{HR} \cdots \geq_{HR} F_n$. If we want to prove that $\bar{H}$ is decreasing in v_1 for fixed $v_2, \ldots, v_n \in (0, 1)$, we choose the following reliability functions:

$$\bar{F}_1(t) = \begin{cases} 1, & \text{for } 0 \leq t \leq 1 \\ 2 - t, & \text{for } 1 < t \leq 2 \\ 0, & \text{for } \quad t > 2 \end{cases}$$

and

$$\bar{F}_i(t) = \begin{cases} 1 - (1 - v_2 \ldots v_i)t, & \text{for } 0 \leq t \leq 1 \\ v_2 \ldots v_i(2 - t), & \text{for } 1 < t \leq 2 \\ 0, & \text{for } \quad t > 2 \end{cases}$$

for $i = 2, \ldots, n$. Hence

$$r_2(t) = \frac{\bar{F}_2(t)}{\bar{F}_1(t)} = \begin{cases} 1 - (1 - v_2)t, & \text{for } 0 \leq t \leq 1 \\ v_2, & \text{for } 1 < t \leq 2 \end{cases}$$

and

$$r_i(t) = \frac{\bar{F}_i(t)}{\bar{F}_{i-1}(t)} = \begin{cases} \frac{1 - (1 - v_2 \ldots v_i)t}{1 - (1 - v_2 \ldots v_{i-1})t}, & \text{for } 0 \leq t \leq 1 \\ v_i, & \text{for } 1 < t \leq 2 \end{cases}$$

for $i = 3, \ldots, n$, which are continuous and decreasing. Therefore, $F_1 \geq_{HR} \cdots \geq_{HR} F_n$ holds and from (3.15), we have that

$$\frac{\bar{Q}_2(\bar{F}_1(t), \ldots, \bar{F}_n(t))}{\bar{Q}_1(\bar{F}_1(t), \ldots, \bar{F}_n(t))} = \bar{H}(2 - t, v_2, \ldots, v_n)$$

is decreasing for $t \in (1, 2)$. So $\bar{H}(v_1, \ldots, v_n)$ is decreasing for $v_1 \in (0, 1)$, for all $v_2, \ldots, v_n \in (0, 1)$. We can prove that it is decreasing in the other variables in a similar way (see Navarro and del Águila 2017). For example, for the second variable, given $v_1, v_3, \ldots, v_n \in (0, 1)$, we can choose the following reliability functions:

$$\bar{F}_1(t) = \begin{cases} 1 - (1 - v_1)t, & \text{for } 0 \leq t \leq 1 \\ v_1, & \text{for } 1 < t \leq 2 \\ v_1(3 - t), & \text{for } 2 < t \leq 3 \\ 0, & \text{for } \quad t > 3 \end{cases}$$

$$\bar{F}_2(t) = \begin{cases} 1 - (1 - v_1)t, & \text{for } 0 \leq t \leq 1 \\ v_1(2 - t), & \text{for } 1 < t \leq 2 \\ 0, & \text{for } \quad t > 2 \end{cases}$$

and

$$\bar{F}_i(t) = \begin{cases} v_i \ldots v_3(1 - (1 - v_1)t), & \text{for } 0 \leq t \leq 1 \\ v_i \ldots v_3 v_1(2 - t), & \text{for } 1 < t \leq 2 \\ 0, & \text{for } \quad t > 2 \end{cases}$$

for $i = 3, \ldots, n$. Hence

$$r_2(t) = \frac{\bar{F}_2(t)}{\bar{F}_1(t)} = \begin{cases} 1, & \text{for } 0 \leq t \leq 1 \\ 2 - t, & \text{for } 1 < t \leq 2 \\ 0, & \text{for } 2 < t \leq 3 \end{cases}$$

and

$$r_i(t) = \frac{\bar{F}_i(t)}{\bar{F}_{i-1}(t)} = \begin{cases} v_i, & \text{for } 0 \leq t \leq 1 \\ v_i, & \text{for } 1 < t \leq 2 \end{cases}$$

for $i = 3, \ldots, n$, and the result holds as above.

The proof of (iii) is similar to the proof of (ii). $\qquad\qquad\square$

Let us see an example which shows how to use the preceding theoretical results to compare systems.

Example 3.5 As in the preceding section, we can consider the series and parallel systems with lifetimes $X_{1:2}$ and $X_{2:2}$, respectively. Now we do not assume a common distribution for the component lifetimes X_1 and X_2. So they have arbitrary distribution functions F_1 and F_2, a copula C and a survival copula $\widehat{C}$. Remember that

$$X_{1:2} \leq_{ST} X_i \leq_{ST} X_{2:2}$$

holds for all F_1, F_2 and all C.

However, if we consider the hazard rate order, then

$$X_{1:2} \leq_{HR} X_1$$

holds for all F_1, F_2 iff $\widehat{C}(u, v)/u$ is increasing in $(0, 1)^2$. Of course, this ordering holds for IND components since $\widehat{C}(u, v)/u = (uv)/u = v$ is increasing (a well known property). In this case, it can be proved that the hazard rate of the series system is $h_{1:2} = h_1 + h_2$, where h_i is the hazard rate of X_i. So $h_{1:2} \geq h_i$ for $i = 1, 2$.

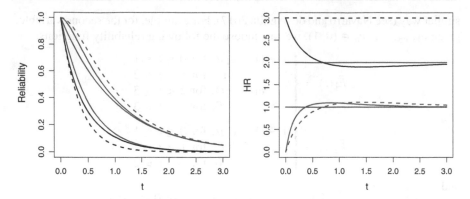

Fig. 3.9 Reliability (left) and hazard rate functions (right) for the series system $X_{1:2}$ (black), the components X_i (red) and the parallel system $X_{2:2}$ (blue) in Example 3.5 for the case of IND components (dashed lines) and dependent (continuous lines) components with the survival copula (3.12)

However, surprisingly, this property is not true when the components are dependent. Thus, if we consider the Clayton–Oakes survival copula (3.12), then

$$\frac{\widehat{C}(u, v)}{u} = \frac{v}{u + v - uv}$$

is decreasing in u and increasing in v. Therefore, for this copula,

$$X_{1:2} \leq_{HR} X_1$$

does not hold for all F_1, F_2. For example, if we consider two exponential distributions $F_i(t) = 1 - \exp(-it)$ for $t \geq 0$ and $i = 1, 2$, then we obtain the reliability (left) and hazard rate (right) functions plotted in Fig. 3.9 for IND components (dashed lines) and dependent components (continuous lines) with the survival copula in (3.12). Note that they are ST ordered in both cases (as expected), that $X_{1:2} \leq_{HR} X_1$ also holds in both cases, that $X_{1:2} \leq_{HR} X_2$ holds for the IND case but that it does not hold for this copula. Note that the used series systems with age t are going to be ST better (i.e., more reliable) than the used components X_2 with the same age t, for $t \geq 0.694$. Also note that they are equivalent when $t \to \infty$. However, the used series systems with age t are going to be ST worse than the used components X_1 with the same age t, for all t.

To explain these properties, we can use Theorem 3.5, (ii) to obtain that $X_{1:2} \leq_{HR} X_1$ holds for all $F_1 \geq_{HR} F_2$ iff the function

$$\bar{H}_1(v_1, v_2) = \frac{\bar{Q}_1(v_1, v_1 v_2)}{\bar{Q}_{1:2}(v_1, v_1 v_2)} = \frac{v_1}{\widehat{C}(v_1, v_1 v_2)}$$

is decreasing in $(0, 1)^2$. Analogously, $X_{1:2} \leq_{HR} X_2$ holds for all $F_1 \geq_{HR} F_2$ iff the function

$$\bar{H}_2(v_1, v_2) = \frac{\bar{Q}_2(v_1, v_1 v_2)}{\bar{Q}_{1:2}(v_1, v_1 v_2)} = \frac{v_1 v_2}{\widehat{C}(v_1, v_1 v_2)}$$

is decreasing in $(0, 1)^2$.

If the components are dependent with the survival copula in (3.12), then

$$\bar{H}_1(v_1, v_2) = \frac{v_1(v_1 + v_1 v_2 - v_1^2 v_2)}{v_1^2 v_2} = \frac{1 + v_2 - v_1 v_2}{v_2},$$

which is decreasing in $(0, 1)^2$, and

$$\bar{H}_2(v_1, v_2) = \frac{v_1 v_2(v_1 + v_1 v_2 - v_1^2 v_2)}{v_1^2 v_2} = 1 + v_2 - v_1 v_2,$$

which is decreasing in v_1 but increasing in v_2. Hence, $X_{1:2} \leq_{HR} X_1$ holds for all $F_1 \geq_{HR} F_2$ (and this copula) but $X_{1:2} \leq_{HR} X_2$ does not hold for all $F_1 \geq_{HR} F_2$ (as we can see in Fig. 3.9). In this case, the series system is HR ordered with the best component (X_1) but not always with the worse one (X_2). If they are ID, both orderings hold (see Fig. 3.5 in the preceding section).

Let us study now the parallel system. For example, $X_1 \leq_{HR} X_{2:2}$ holds for all F_1, F_2 iff

$$\frac{u + v - \widehat{C}(u, v)}{u} = 1 + \frac{v - \widehat{C}(u, v)}{u}$$

is decreasing in $(0, 1)^2$. If the components are IND, then

$$\frac{v - \widehat{C}(u, v)}{u} = v\left(\frac{1}{u} - 1\right)$$

which is increasing in v and decreasing in u. So, surprisingly, this ordering does not hold for all F_1, F_2 even if the components are IND, as can be seen in Fig. 3.9, right, where $X_{2:2}$ (dashed blue line) and the best component X_1 (bottom red line) are not HR ordered. However, $X_{2:2}$ (dashed blue line) and the worse component X_2 (top red line) are HR ordered.

In this figure, the same holds for the Clayton–Oakes copula. As above we can use Theorem 3.5, (ii), to study if this is a general property. Thus $X_1 \leq_{HR} X_{2:2}$ holds for all $F_1 \geq_{HR} F_2$ iff the function

$$\bar{H}_3(v_1, v_2) = \frac{\bar{Q}_{2:2}(v_1, v_1 v_2)}{\bar{Q}_1(v_1, v_1 v_2)} = \frac{v_1 + v_1 v_2 - \widehat{C}(v_1, v_1 v_2)}{v_1}$$

is decreasing in $(0, 1)^2$. Analogously, $X_2 \leq_{HR} X_{2:2}$ holds for all $F_1 \geq_{HR} F_2$ iff the function

$$\bar{H}_4(v_1, v_2) = \frac{\bar{Q}_{2:2}(v_1, v_1 v_2)}{\bar{Q}_2(v_1, v_1 v_2)} = \frac{v_1 + v_1 v_2 - \widehat{C}(v_1, v_1 v_2)}{v_1 v_2}$$

is decreasing in $(0, 1)^2$. If the components are IND, then

$$\bar{H}_3(v_1, v_2) = \frac{v_1 + v_1 v_2 - v_1^2 v_2}{v_1} = 1 + v_2 - v_1 v_2$$

which is decreasing in v_1 and increasing in v_2. So X_1 and $X_{2:2}$ are not HR ordered in Fig. 3.9. However, if the components are IND, then

$$\bar{H}_4(v_1, v_2) = \frac{v_1 + v_1 v_2 - v_1^2 v_2}{v_1 v_2} = \frac{1}{v_2} + 1 - v_1$$

which is decreasing in both v_1 and v_2. So X_2 and $X_{2:2}$ are HR ordered in Fig. 3.9. This is a general property for IND ordered components (the parallel system is HR better that the worse component). This property also holds for the chosen Clayton–Oakes copula since

$$\bar{H}_4(v_1, v_2) = 1 + \frac{1 - v_1 v_2}{v_2(1 + v_2 - v_1 v_2)}$$

is decreasing in both v_1 and v_2. However, X_1 and $X_{2:2}$ are not HR ordered as can be seen in Fig. 3.9, right. Also note that the series systems in both cases are HR ordered but that the parallel systems are not. The code in R to get these plots is the following:

```
#Reliability functions:
#IID case:
R1<-function(t) exp(-t)
R2<-function(t) exp(-2*t)
QIND<-function(u,v) u*v
G12<-function(t) QIND(R1(t),R2(t))
G22<-function(t) R1(t)+R2(t)-G12(t)
curve(G12(x),xlab='t',ylab='Reliability',0,3,lty=2,lwd=2)
curve(G22(x),add=T,col='blue',lty=2,lwd=2)
curve(R1(x),add=T,col='red',lty=2,lwd=2)
curve(R2(x),add=T,col='red',lty=2,lwd=2)
#Clayton
C<-function(u,v) u*v/(u+v-u*v)
R12<-function(t) C(R1(t),R2(t))
R22<-function(t) R1(t)+R2(t)-R12(t)
curve(R12(x),xlab='t',add=T,lwd=2)
curve(R22(x),add=T,col='blue',lwd=2)

#Hazard rate functions
#IND case
f1<-function(t) exp(-t)
f2<-function(t) 2*exp(-2*t)
Q1IID<-function(u,v) v #partial derivative 1
Q2IID<-function(u,v) u #partial derivative2
g12<-function(t) {
   f1(t)*Q1IID(R1(t),R2(t))+f2(t)*Q2IID(R1(t),R2(t))
}
g22<-function(t) f1(t)+f2(t)-g12(t)
curve(g12(x)/G12(x),0,3,ylab='HR',ylim=c(0,3),lty=2,lwd=2)
curve(g22(x)/G22(x),add=T,col='blue',lty=2,lwd=2)
curve(f1(x)/R1(x),add=T,col='red',lwd=2)
curve(f2(x)/R2(x),add=T,col='red',lwd=2)
#Clayton
C1<-function(u,v) v^2/(u+v-u*v)^2 #partial derivative 1
```

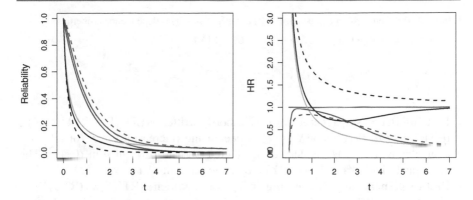

Fig. 3.10 Reliability (left) and hazard rate functions (right) for the series system $X_{1:2}$ (black), the parallel system $X_{2:2}$ (blue) and the components X_1 (Exponential, red) and X_2 (Pareto, green) in Example 3.5 for the case of IND components (dashed lines) and dependent components (continuous lines) with the survival copula (3.12)

```
C2<-function(u,v) u^2/(u+v-u*v)^2 #partial derivative 2
f12<-function(t) f1(t)*C1(R1(t),R2(t))+f2(t)*C2(R1(t),R2(t))
f22<-function(t) f1(t)+f2(t)-f12(t)
curve(f12(x)/R12(x),add=T,lwd=2)
curve(f22(x)/R22(x),add=T,col='blue',lwd=2)
```

We can modify this code to plot these functions for other marginals and/or other copulas. For example, if we consider the same exponential for X_1 but the Pareto distribution $F_2(t) = 1 - 1/(1 + 5t)$ for $t \geq 0$ for the second component lifetime X_2, then they are not ordered and we obtain the plot in Fig. 3.10 (for the same copula). Note that the series and parallel systems are HR ordered in the case of IND components (dashed lines) but that they are not ordered for the Clayton–Oakes copula (black and blue continuous lines). This is a really surprising property!   ◀

As in the preceding section, we can obtain conditions for distribution-free ordering results based on properties of the copula and/or the survival copula. These conditions are related with negative dependence properties. These relationships were studied in Navarro et al. (2021). Let us see some examples. The proofs are straightforward.

To get these results we need the definitions of well-known dependence properties and how they can be stated in terms of copulas. A continuous random pair (X, Y) with copula C is said to be:

- **Positive (Negative) Quadrant Dependent**, shortly written as PQD (NQD), iff $\Pr(X \leq x, Y \leq y) \geq \Pr(X \leq x)\Pr(Y \leq y)$ for all x, y. If the marginal distributions are continuous, then the PQD (NQD) property is equivalent (see Proposition 3.9) to $C(u, v) \geq uv$ ($C(u, v) \leq uv$) in $[0, 1]^2$;

Table 3.2 Relationships among positive (left) and negative (right) dependence properties

$$SI(Y|X) \ \Rightarrow LTD(Y|X) \quad SD(Y|X) \ \Rightarrow LTI(Y|X)$$
$$\Downarrow \qquad\qquad \Downarrow \qquad\qquad \Downarrow \qquad\qquad \Downarrow$$
$$RTI(Y|X) \Rightarrow \quad PQD \qquad RTD(Y|X) \Rightarrow \quad NQD$$

- **Left Tail Decreasing (Increasing)** in X, shortly written as $LTD(Y|X)$ ($LTI(Y|X)$), if, and only if, $\Pr(Y \leq y|X \leq x)$ is decreasing (increasing) in x for all y or, equivalently, $C(u, v)/u$ is decreasing (increasing) in u for all v in $(0, 1)^2$. The concepts $LTD(X|Y)$ and $LTI(X|Y)$ are defined in a similar way;
- **Right Tail Increasing (Decreasing)** in X, shortly written as $RTI(Y|X)$ ($RTD(Y|X)$), if, and only if, $\Pr(Y > y|X > x)$ is increasing (decreasing) in x for all y or, equivalently, $\widehat{C}(u, v)/u$ is decreasing (increasing) in u for all v in $(0, 1)^2$;
- **Stochastically Increasing (Decreasing)** in X, shortly written as $SI(Y|X)$ ($SD(Y|X)$), if, and only if, $(Y|X = x)$ is ST-increasing (decreasing) in x.

We say that (X, Y) is LTD if it is both $LTD(Y|X)$ and $LTD(X|Y)$. The concepts LTI, RTI, RTD, SI and SD are defined similarly. The relationships among the above dependence properties are summarized in Table 3.2. Also note that the PQD (NQD) property implies that the Pearson correlation, Spearman correlation and Kendal tau coefficients are nonnegative (nonpositive), see Nelsen (2006). So all of them are positive (negative) dependence properties.

In the first proposition we compare the components with the series system.

Proposition 3.11 *Let X_1 and X_2 be component lifetimes with survival copula $\widehat{C}$ and distribution functions F_1 and F_2, respectively. Then the following statements are equivalent:*

(i) *$X_{1:2} \leq_{HR} X_1$ holds for all F_1 and F_2;*
(ii) *$\widehat{C}(u, v)/u$ is increasing in $u \in (0, 1)$ for every $v \in (0, 1)$;*
(iii) *$(v - 1 + C(1 - u, 1 - v))/u$ is increasing in $u \in (0, 1)$ for every $v \in (0, 1)$;*
(iv) *(X_1, X_2) is $RTD(X_2|X_1)$.*

Note that we need a negative dependence property (RTD) to separate the series system from its components. For the RHR order we get the following conditions.

Proposition 3.12 *Let X_1 and X_2 be component lifetimes with copula C and with distribution functions F_1 and F_2, respectively. Then the following statements are equivalent:*

(i) *$X_1 \leq_{RHR} X_{2:2}$ holds for all F_1 and F_2;*
(ii) *$C(u, v)/u$ is increasing in $u \in (0, 1)$ for every $v \in (0, 1)$;*
(iii) *$(v - 1 + \widehat{C}(1 - u, 1 - v))/u$ is increasing in $u \in (0, 1)$ for every $v \in (0, 1)$.*
(iv) *(X_1, X_2) is $LTI(X_2|X_1)$.*

Note that here we also need a negative dependence property and that the conditions are duals (by changing $\widehat{C}$ with C). To compare series and parallel systems we have the following condition.

Proposition 3.13 *Let X_1 and X_2 be component lifetimes with survival copula $\widehat{C}$ and with distribution functions F_1 and F_2, respectively. Then:*

(i) $X_{1:2} \leq_{HR} X_{2:2}$ *holds for all F_1, F_2 iff $\widehat{C}(u, v)/(u + v)$ is increasing in $(0, 1)^2$;*
(ii) $X_{1:2} \leq_{RHR} X_{2:2}$ *holds for all F_1, F_2 iff $C(u, v)/(u + v)$ is increasing in $(0, 1)^2$.*

For ordered components we have the following results.

Proposition 3.14 *Let X_1 and X_2 be component lifetimes. Then:*

(i) $X_{1:2} \leq_{HR} X_1$ *holds for all $F_1 \geq_{HR} F_2$ iff $\widehat{C}(u, uv)/u$ is increasing in $(0, 1)^2$;*
(ii) $X_{1:2} \leq_{HR} X_2$ *holds for all $F_1 \geq_{HR} F_2$ iff $\widehat{C}(u, uv)/(uv)$ is increasing in $(0, 1)^2$;*
(iii) $X_1 \leq_{HR} X_{2:2}$ *holds for all $F_1 \geq_{HR} F_2$ iff $(uv - \widehat{C}(u, uv))/u$ is decreasing in $(0, 1)^2$;*
(iv) $X_2 \leq_{HR} X_{2:2}$ *holds for all $F_1 \geq_{HR} F_2$ iff $(u - \widehat{C}(u, uv))/(uv)$ is decreasing in $(0, 1)^2$;*
(v) $X_{1:2} \leq_{HR} X_{2:2}$ *holds for all $F_1 \geq_{HR} F_2$ iff $\widehat{C}(u, uv)/(u + uv)$ is increasing in $(0, 1)^2$.*

Note that all the conditions for the survival copula $\widehat{C}$ in the preceding proposition can be seen as negative dependence properties.

Proposition 10 in Navarro et al. (2021) proves that, for any copula C, X_1 and $X_{2:2}$ are not HR ordered for all F_1, F_2. Note that $X_1 \leq_{HR} X_{2:2}$ holds for all F_1, F_2 iff

$$\frac{1 - C(1 - u, 1 - v)}{u}$$

is decreasing in $(0, 1)^2$. However, note that this ratio is always increasing in v. Hence, the results given in Example 3.5 for X_1 and $X_{2:2}$ are valid for any copula C (i.e. for some distribution functions F_1 and F_2 they are not HR ordered).

Analogously, it can be proved that X_1 and $X_{1:2}$ are not RHR ordered for all F_1, F_2. To get these orderings we need to assume ordered components (as stated in the preceding proposition).

As in the preceding section, we can compare systems with dependent and independent components. If X_1 and X_2 have a copula C and a survival copula $\widehat{C}$ and Y_1 and Y_2 are independent, X_1 and Y_1 have the common distribution function F_1 and X_2 and Y_2 have the common distribution function F_2, then we obtain the following results.

Proposition 3.15 *The following statements are equivalent:*

 (i) $X_{1:2} \geq_{ST} Y_{1:2}$ (respectively, $\leq_{ST}$) holds for all F_1 and F_2;
 (ii) $X_{2:2} \leq_{ST} Y_{2:2}$ (respectively, $\geq_{ST}$) holds for all F_1 and F_2;
 (iii) $C(u, v) \geq uv$ (respectively, $C(u, v) \leq uv$) in $[0, 1]^2$;
 (iv) $\widehat{C}(u, v) \geq uv$ (respectively, $\widehat{C}(u, v) \leq uv$) in $[0, 1]^2$;
 (v) (X_1, X_2) is PQD (respectively, NQD).

Proposition 3.16 *The following statements are equivalent:*

 (i) $X_{1:2} \geq_{HR} Y_{1:2}$ (respectively, $\leq_{HR}$) holds for all F_1 and F_2;
 (ii) $\widehat{C}(u, v)/(uv)$ is decreasing (respectively, increasing) in $(0, 1)^2$;
 (iii) (X_1, X_2) is RTI (respectively, RTD).

Proposition 3.17 *The following statements are equivalent:*

 (i) $X_{2:2} \geq_{RHR} Y_{2:2}$ (respectively, $\leq_{RHR}$) holds for all F_1 and F_2;
 (ii) $C(u, v)/(uv)$ is increasing (respectively, decreasing) in $(0, 1)^2$;
 (iii) (X_1, X_2) is LTI (respectively, LTD).

As we have seen in Example 3.5, the comparison results for the general case can also be applied to systems with IND components. The results for all the semi-coherent systems with 1-3 components were obtained in Navarro and del Águila (2017). Their dual distortion functions are given in Table 3.3. All the ST and HR orderings for these

Table 3.3   Dual distortions functions of coherent systems with 1–3 independent components

N	$T = \psi(X_1, X_2, X_3)$	$\overline{Q}(u_1, u_2, u_3)$
1	$X_{1:3} = \min(X_1, X_2, X_3)$	$u_1 u_2 u_3$
2	$\min(X_2, X_3)$	$u_2 u_3$
3	$\min(X_1, X_3)$	$u_1 u_3$
4	$\min(X_1, X_2)$	$u_1 u_2$
5	$\min(X_3, \max(X_1, X_2))$	$u_1 u_3 + u_2 u_3 - u_1 u_2 u_3$
6	$\min(X_2, \max(X_1, X_3))$	$u_1 u_2 + u_2 u_3 - u_1 u_2 u_3$
7	$\min(X_1, \max(X_2, X_3))$	$u_1 u_2 + u_1 u_3 - u_1 u_2 u_3$
8	X_3	u_3
9	X_2	u_2
10	X_1	u_1
11	$X_{2:3}$	$u_1 u_2 + u_1 u_3 + u_2 u_3 - 2u_1 u_2 u_3$
12	$\max(X_3, \min(X_1, X_2))$	$u_3 + u_1 u_2 - u_1 u_2 u_3$
13	$\max(X_2, \min(X_1, X_3))$	$u_2 + u_1 u_3 - u_1 u_2 u_3$
14	$\max(X_1, \min(X_2, X_3))$	$u_1 + u_2 u_3 - u_1 u_2 u_3$
15	$\max(X_2, X_3)$	$u_2 + u_3 - u_2 u_3$
16	$\max(X_1, X_3)$	$u_1 + u_3 - u_1 u_3$
17	$\max(X_1, X_2)$	$u_1 + u_2 - u_1 u_2$
18	$X_{3:3} = \max(X_1, X_2, X_3)$	$u_1 + u_2 + u_3 - u_1 u_2 - u_1 u_3 - u_2 u_3 + u_1 u_2 u_3$

Table 3.4 Relationships for the ST order between the coherent systems with independent components given in Table 3.3. The value 2 indicates that $T_i \leq_{ST} T_j$ holds for any F_1, F_2, F_3 (i denotes the row and j the column). The value 1 indicates that $T_i \leq_{ST} T_j$ holds for all $F_1 \geq_{ST} F_2 \geq_{ST} F_3$. It also indicates that $T_i \leq_{ST} T_j$ does not hold for all F_1, F_2, F_3. The value 0 indicates that $T_i \leq_{ST} T_j$ does not hold for all $F_1 \geq_{ST} F_2 \geq_{ST} F_3$

ST	2	3	4	5	6	7	8	9	10	11	12	13	14	15	16	17	18
1	2	2	2	2	2	2	2	2	2	2	2	2	2	2	2	2	2
2	2	1	1	2	2	1	2	2	1	2	2	2	2	2	2	2	2
3	0	2	1	2	1	2	2	1	2	2	2	2	2	2	2	2	2
4	0	0	2	0	2	2	0	2	2	2	2	2	2	2	2	2	2
5	0	0	0	2	1	1	2	1	1	2	2	2	2	2	2	2	2
6	0	0	0	0	2	1	0	2	1	2	2	2	2	2	2	2	2
7	0	0	0	0	0	2	0	0	2	2	2	2	2	2	2	2	2
8	0	0	0	0	0	0	2	1	1	0	2	1	1	2	2	1	2
9	0	0	0	0	0	0	0	2	1	0	0	2	1	2	1	2	2
10	0	0	0	0	0	0	0	0	2	0	0	0	2	0	2	2	2
11	0	0	0	0	0	0	0	0	0	2	2	2	2	2	2	2	2
12	0	0	0	0	0	0	0	0	0	0	2	1	1	2	2	1	2
13	0	0	0	0	0	0	0	0	0	0	0	2	1	2	1	2	2
14	0	0	0	0	0	0	0	0	0	0	0	0	2	0	2	2	2
15	0	0	0	0	0	0	0	0	0	0	0	0	0	2	1	1	2
16	0	0	0	0	0	0	0	0	0	0	0	0	0	0	2	1	2
17	0	0	0	0	0	0	0	0	0	0	0	0	0	0	0	2	2
18	0	0	0	0	0	0	0	0	0	0	0	0	0	0	0	0	2

systems are given in Tables 3.4 and 3.5. The value 2 indicates that the ordering holds for any components, the value 1 that it only holds for ordered components and the value 0 that it does not hold for ordered components. The relationships for the HR order and ordered components are summarized in the graph given in Fig. 3.11.

We conclude this section by showing how to proceed when the systems are built just by using two kind of components. Here we assume that T_1 and T_2 are the lifetimes of two coherent systems with components having one of the two (different) distribution functions F (type A) or G (type B). For example, we can consider the systems in Fig. 3.12.

Under this assumption, it is clear that the system reliability functions can be written as

$$\bar{F}_{T_i}(t) = \bar{Q}_i(\bar{F}, \bar{G})$$

for $i = 1, 2$, where $\bar{Q}_i : [0, 1]^2 \to [0, 1]$ are two (bivariate) distortion functions. They can be obtained from the general distortion functions obtained in Chap. 2 (by using minimal path or cut sets). Under some exchangeabillity assumptions between the components of the same type, these distortion functions can also be computed

Table 3.5 Relationships for the HR order between the coherent systems with independent components given in Table 3.3. The value 2 indicates that $T_i \leq_{HR} T_j$ holds for any F_1, F_2, F_3 (i denotes the row and j the column). The value 1 indicates that $T_i \leq_{HR} T_j$ holds for all $F_1 \geq_{HR} F_2 \geq_{HR} F_3$. It also indicates that $T_i \leq_{HR} T_j$ does not hold for all F_1, F_2, F_3. The value 0 means that $T_i \leq_{HR} T_j$ does not hold for all $F_1 \geq_{HR} F_2 \geq_{HR} F_3$

HR	2	3	4	5	6	7	8	9	10	11	12	13	14	15	16	17	18
1	2	2	2	2	2	2	2	2	2	2	2	2	2	2	2	2	2
2	2	1	1	1	1	1	2	2	1	1	1	1	1	2	1	1	1
3	0	2	1	0	0	1	2	1	2	0	1	1	1	1	2	1	1
4	0	0	2	0	0	0	0	2	2	0	0	0	0	0	0	2	0
5	0	0	0	2	0	0	2	1	1	0	0	1	1	1	1	2	2
6	0	0	0	0	2	0	0	2	1	0	0	0	1	0	2	1	2
7	0	0	0	0	0	2	0	0	2	0	0	0	1	2	1	1	2
8	0	0	0	0	0	0	2	1	1	0	0	0	0	1	1	1	1
9	0	0	0	0	0	0	0	2	1	0	0	0	0	0	0	1	0
10	0	0	0	0	0	0	0	0	2	0	0	0	0	0	0	0	0
11	0	0	0	0	0	0	0	0	0	2	0	1	1	2	2	2	2
12	0	0	0	0	0	0	0	0	0	0	2	0	1	1	1	1	1
13	0	0	0	0	0	0	0	0	0	0	0	2	1	0	0	1	0
14	0	0	0	0	0	0	0	0	0	0	0	0	2	0	0	0	0
15	0	0	0	0	0	0	0	0	0	0	0	0	0	2	1	1	1
16	0	0	0	0	0	0	0	0	0	0	0	0	0	0	2	0	0
17	0	0	0	0	0	0	0	0	0	0	0	0	0	0	0	2	0
18	0	0	0	0	0	0	0	0	0	0	0	0	0	0	0	0	2

from the **survival signature** defined in Coolen and Coolen-Maturi (2012) (see also Samaniego and Navarro 2016).

For example, for the systems in Fig. 3.12, if we assume IND components, we have

$$\bar{Q}_1(u, v) = uv + v^2 - uv^2$$

and

$$\bar{Q}_2(u, v) = 2uv - uv^2$$

for $u, v \in [0, 1]^2$.

In this case $T_1 \leq_{ST} T_2$ (resp. $\geq_{ST}$) holds for all F, G iff $\bar{Q}_1 \leq \bar{Q}_2$ (resp. $\geq_{ST}$). If we define the **difference function**

$$\Delta(u, v) := \bar{Q}_2(u, v) - \bar{Q}_1(u, v),$$

this ordering holds for all F, G iff $\Delta(u, v) \geq 0$ (resp. ≤ 0) for all $u, v \in [0, 1]$.

However, in some cases, we need conditions between F and G to get this ordering. Thus, for the systems in Fig. 3.12, we obtain

$$\Delta(u, v) = 2uv - uv^2 - (uv + v^2 - uv^2) = uv - v^2.$$

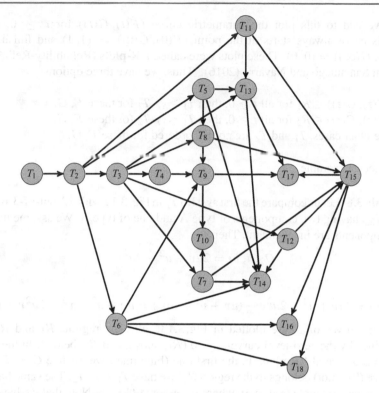

Fig. 3.11 Hazard rate ordering relationships between the coherent systems with 1–3 independent components given in Table 3.3 when $F_1 \geq_{HR} F_2 \geq_{HR} F_3$ holds

Fig. 3.12 Two coherent systems of order 3 with a similar structure built with components of type A and B.

Therefore, $\Delta(u, v) \geq 0 \ (\leq 0)$ iff $u \geq v \ (\leq)$. Hence, $T_1 \leq_{ST} T_2$ (resp. $\geq_{ST}$) holds iff $\bar{F} \geq \bar{G} \ (\leq)$, that is, the best component should be placed at the first position (as expected). Note that they are not ST ordered when F and G are not ST ordered.

In other situations, the conditions to get this ordering can be more complicated. In this case we can proceed as follows. We plot the level curves (contour plot) of Δ in $[0, 1]^2$. In this plot we highlight the border line which leads to $\Delta = 0$ and we define the regions

$$\mathcal{R}_1 = \{(u, v) \in [0, 1] : \Delta(u, v) \leq 0\}$$

and

$$\mathcal{R}_2 = \{(u, v) \in [0, 1] : \Delta(u, v) \geq 0\}.$$

Then we add to this plot the parametric curve $(\bar{F}(t), \bar{G}(t))$ for $t \geq 0$. Note that this curve always starts at the point $(\bar{F}(0), \bar{G}(0)) = (1, 1)$ and finished at $(\bar{F}(\infty), \bar{G}(\infty)) = (0, 0)$. These plots were called **RR-plots** (Reliability-Reliability plots) in Samaniego and Navarro (2016). Thus, we have three options:

- If $(\bar{F}(t), \bar{G}(t)) \in \mathcal{R}_1$ for all $t \geq 0$, then $T_1 \geq_{ST} T_2$ for these F, G.
- If $(\bar{F}(t), \bar{G}(t)) \in \mathcal{R}_2$ for all $t \geq 0$, then $T_1 \leq_{ST} T_2$ for these F, G.
- In the other cases, T_1 and T_2 are not ST ordered for these F, G.

Let us see an example.

Example 3.6 Let us compare the first system T_1 in Fig. 3.12 with a 2-out-of-3 system $T_2 = X_{2:3}$ having two components of type A and one of type B. We assume that all the components are independent. Then

$$\bar{Q}_2(u, v) = 2uv + u^2 - 2u^2v$$

and

$$\Delta(u, v) = 2uv + u^2 - 2u^2v - (uv + v^2 - uv^2) = uv + u^2 - v^2 - 2u^2v + uv^2.$$

The level curves of Δ are plotted in Fig. 3.13, left. The regions $\mathcal{R}_1$ and $\mathcal{R}_2$ are determined by the zero-level curve $\Delta = 0$ ($\mathcal{R}_1$, above, and $\mathcal{R}_2$, below). In the right plot we add several RR-plots. In the first one (blue line), we assume $\bar{G} = \bar{F}^2$. As the curve (RR-plot) belongs to the region $\mathcal{R}_2$, we have $T_1 \leq_{ST} T_2$. The same happen for the second example (red line), where we assume $\bar{G} = \bar{F}$. Note that we have this property $T_1 \leq_{ST} T_2$ for all $\bar{G} \leq \bar{F}$ (and also for some $\bar{G} \geq \bar{F}$). However, in the third case (green line), we assume $\bar{G}^2 = \bar{F}$ and the curve crosses both regions. Therefore, T_1 and T_2 are not ST-ordered. Finally, we choose $G = F^3$ (i.e. $\bar{G} = 1 - (1 - \bar{F})^3$) and then the curve (purple line) belongs to the region $\mathcal{R}_1$. So we have $T_1 \geq_{ST} T_2$. Note that the level curves can be used to determine approximately the difference

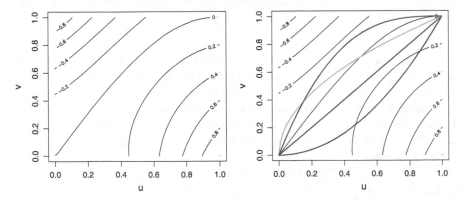

Fig. 3.13 Level curves of Δ for the systems in Example 3.6 and RR-plots (right) when we assume $\bar{G} = \bar{F}^2$ (blue line), $\bar{G} = \bar{F}$ (red line), $\bar{G}^2 = \bar{F}$ (green line) and $G = F^3$ (purple line)

between both system reliability functions. For example, if $F = G$ (red line), then $0 \leq \bar{F}_2 - \bar{F}_1 \leq 0.2$. The R code to get these plots is the following:

```
# RR-plots
Q1<-function(u,v)  u*v+v^2-u*v^2
Q2<-function(u,v)  2*u*v+u^2-2*u^2*v
D<-function(u,v)   Q2(u,v)-Q1(u,v)
x<-seq(0,1,0.01)
y<-seq(0,1,0.01)
z<-outer(x,y,D)
contour(x,y,z,xlab='u',ylab='v')
curve(x^ 2,add=T,col='blue',lwd=2)
curve(x+1-1,add=T,col='red',lwd=2)
curve(x^0.5,add=T,col='green',lwd=2)
curve(1-(1-x)^3,add=T,col='purple',lwd=2)        ◄
```

3.5 A Parrondo Paradox in Reliability

The Parrondo's paradox shows how, in some games, a random strategy might be better than any deterministic strategy. Di Crescenzo (2007) noted that a similar paradox holds in reliability for series systems with independent heterogeneous components. The problem can be stated as follows.

Let $T = \min(X_1, X_2)$ be the lifetime of a series system with two independent components having reliability functions $\bar{F}_1$ and $\bar{F}_2$. We can assume that the components of type 1 are better than the others, that is, $\bar{F}_1 \geq \bar{F}_2$ (but we will see later that we do not need this assumption).

On the other hand, we can consider the series system with lifetime $S = \min(Y_1, Y_2)$, where Y_1 and Y_2 are IID with common reliability

$$\bar{G} = \frac{1}{2}\bar{F}_1 + \frac{1}{2}\bar{F}_2.$$

This system represents the case in which we choose the components randomly from a mixed population with a 50% of units of type 1 (with reliability $\bar{F}_1$) and a 50% of units of type 2 (with reliability $\bar{F}_2$), while in the first option we choose for sure one component of each type.

Which one is the best option? Does this property depend on $\bar{F}_1$ and $\bar{F}_2$? What is the best general option? Could this property be extended to other system structures? What happen if the components are dependent?

The respective system reliability functions in both options can be represented with distortions as

$$\bar{F}_T(t) = \Pr(X_1 > t, X_2 > t) = \bar{F}_1(t)\bar{F}_2(t) = \bar{Q}_T(\bar{F}_1(t), \bar{F}_2(t))$$

and

$$\bar{F}_S(t) = \Pr(Y_1 > t, Y_2 > t) = \bar{G}(t)\bar{G}(t) = \left(\frac{1}{2}\bar{F}_1(t) + \frac{1}{2}\bar{F}_2(t)\right)^2 = \bar{Q}_S(\bar{F}_1(t), \bar{F}_2(t))$$

with

$$\bar{Q}_T(u_1, u_2) = u_1 u_2$$

and

$$\bar{Q}_S(u_1, u_2) = \left(\frac{u_1 + u_2}{2}\right)^2$$

for $u_1, u_2 \in [0, 1]$. It is easy to prove that $\bar{Q}_T \leq \bar{Q}_S$ since

$$\sqrt{u_1 u_2} \leq \frac{u_1 + u_2}{2}$$

(the geometric mean is always less than the arithmetic mean), or just since

$$4u_1 u_2 \leq u_1^2 + 2u_1 u_2 + u_2^2$$

holds for all $u_1, u_2 \in [0, 1]$ because $0 \leq (u_1 - u_2)^2$. Note that we do not need the condition $u_1 \geq u_2$, that is, $\bar{F}_1 \geq \bar{F}_2$. They can be ordered in the reverse sense or even not ordered. In any case, the system with randomly chosen components is always ST better, that is, $T \leq_{ST} S$ for all $\bar{F}_1, \bar{F}_2$. So the Parrondo paradox holds!

The respective reliability functions for exponential components with means 5 and 1 can be seen in Fig. 3.14, left. In the right plot the first unit has a Weibull distribution with reliability $\bar{F}_1(t) = \exp(-t^4)$ for $t \geq 0$. The reliability functions of T and S are plotted in black and blue, respectively. Note that the first one is always worse than the second. As mentioned above, this property holds for all $\bar{F}_1, \bar{F}_2$. The red and orange plots correspond to the series systems obtained with just units of type 1 (red) or 2 (orange). Of course, in the left plot, the best option is the red curve, that is, the series system obtained with just the best units. In many situations this is not a

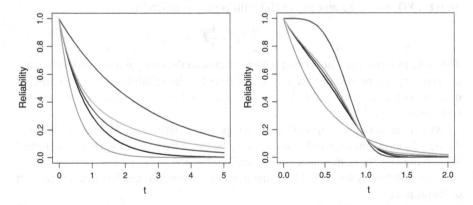

Fig. 3.14 Reliability functions for the series systems T (black) and S (blue) in Parrondo paradox for exponential (left) and Weibull (right) distributions. The red and orange plots correspond to the series systems obtained with just units of type 1 (red) or 2 (orange)

realistic option since we do not use the units of type 2, the bests units could be more expensive, we mighty not know which ones are the best units, or the units could be not ordered (as in the right plot). However, in the green plot we use an 50% of units of type 1 and 2 obtaining a better series system. How can we build this system? To answer this question we need some additional results.

First we are going to study when this "Parrondo paradox" holds. It is easy to see that it can be extended to series systems with n independent components. The explanation is simple since these systems are better when the units are similar (homogeneous). Hence, here the Parrondo paradox is not a paradox but an expectable property. This property is reverted for parallel systems since, in this case, the systems are better when the units are different (heterogeneous). What happen in other system structures? Do these properties hold when the components are dependent?

The answers to some of these questions were obtained in Navarro and Spizzichino (2010). They are based on the notions of Schur-concave and weakly Schur-concave functions defined as follows (see Durante and Papini 2007).

Definition 3.9 A function $g : \mathbb{R}^n \to \mathbb{R}$ is weakly Schur-concave (convex) if

$$g(u_1, \ldots, u_n) \leq g(\bar{u}, \ldots, \bar{u}) \; (\geq)$$

for all $u_1, \ldots, u_n$, where $\bar{u} = (u_1 + \cdots + u_n)/n$.

Definition 3.10 A function $g : \mathbb{R}^n \to \mathbb{R}$ is Schur-concave (convex) if

$$g(u_1, \ldots, u_n) \leq g(v_1, \ldots, v_n) \; (\geq)$$

for all $u_1, \ldots, u_n, v_1, \ldots, v_n$ such that $u_1 + \cdots + u_n = v_1 + \cdots + v_n$ and such that

$$\sum_{i=1}^{j} u_{i:n} \leq \sum_{i=1}^{j} v_{i:n}$$

for all $j = 1, \ldots, n - 1$, where $u_{i:n}$ and $v_{i:n}$ are the ordered values obtained from the respective vectors.

To explain the meaning of these properties let us consider $n = 2$. In both cases, we study the monotonicity of function $g(u_1, u_2)$ when we move the points in the line $u_1 + u_2 = c$. The function g is Schur-concave when it is increasing when the points move to the diagonal. Obviously, then the maximum value is obtained in the point at the diagonal $(\bar{u}, \bar{u})$, that is, then it is also weakly Schur-concave. For example, the function $g(u_1, u_2) = u_1 u_2$ is Schur-concave since if we assume $u_1 + u_2 = c$, then

$$g(u_1, u_2) = u_1 u_2 = u_1(c - u_1)$$

which is increasing for $u_1 \leq c/2$ and decreasing for $u_1 \geq c/2$. Its maximum value is obtained when $u_1 = c/2$, that is, $u_1 = u_2$. The 3D plot and contour plot (level curves) can be seen in Fig. 3.15. Note that g increases when we move to the diagonal (mountain shape). Analogously, it can be proved that $g(u_1, u_2) = 1 - (1 - u_1)(1 - u_2)$ is Schur-convex. The code for these plots is the following:

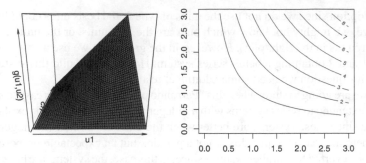

Fig. 3.15 Plot (left) and contour plot (right) for $g(u_1, u_2) = u_1 u_2$

```
#Schur-concave
g<-function(x,y) x*y
x<-seq(0,3,length=50)
y<-seq(0,3,length=50)
z<-outer(x,y,g)
persp(x,y,z,xlab='u1',ylab='u2',zlab='g(u1,u2)',col='red')
contour(x,y,z,col='blue')
```

Now we can state the following result.

Theorem 3.6 (Navarro and Spizzichino 2010) *Let $\bar{Q}$ be the dual distortion function of a system. Then the Parrondo paradox holds (is reverted) for this system if and only if $\bar{Q}$ is weakly Schur-concave (convex).*

Proof Note that to check the Parrondo paradox we have to compare

$$\bar{F}_T(t) = \bar{Q}(\bar{F}_1(t), \dots, \bar{F}_n(t))$$

with

$$\bar{F}_S(t) = \bar{Q}(\bar{G}(t), \dots, \bar{G}(t))$$

where $\bar{G} = (\bar{F}_1 + \dots + \bar{F}_n)/n$. Hence, $\bar{F}_T \leq \bar{F}_S$ holds if and only if $\bar{Q}$ is weakly Schur-concave. The property is reverted when $\bar{Q}$ is weakly Schur-convex. □

Of course, in particular, the Parrondo paradox holds (is reverted) when $\bar{Q}$ is Schur-concave (convex). For series systems with independent components, we have

$$\bar{Q}_{1:n}(u_1, \dots, u_n) = u_1 \dots u_n,$$

which is Schur-concave. So the Parrondo paradox holds for any $F_1, \dots, F_n$. If the components are dependent with a survival copula $\widehat{C}$, then

$$\bar{Q}_{1:n}(u_1, \dots, u_n) = \widehat{C}(u_1, \dots, u_n).$$

Hence, the Parrodo paradox holds if and only if $\widehat{C}$ is weakly Schur-concave. Many copulas are Schur-concave (see Nelsen 2006). For example, all the Archimedean copulas are Schur-concave. Do not exist strict Schur-convex copulas. There are some copulas that are at the same time Schur-convex and Schur-concave (i.e. they are Schur-constant). For them, both option coincide (i.e. $T =_{ST} S$).

However, Durante and Papini (2007) obtained a strict weakly Schur-convex copula. Hence, under this survival copula, the Parrondo paradox is reverted in this series system with dependent components. This is really a paradox since, in this case, it is better to have heterogeneous components in a series system!

These properties are reverted for parallel systems. If the components are independent, then their dual distortion function is

$$\bar{Q}_{n:n}(u_1, \ldots, u_n) = 1 - (1 - u_1) \ldots (1 - u_n),$$

which is Schur-convex in $[0, 1]^n$ and so the Parrondo paradox is reverted. If the components are dependent with a copula C, then

$$\bar{Q}_{n:n}(u_1, \ldots, u_n) = 1 - C(1 - u_1, \ldots, 1 - u_n)$$

and so the Parrondo paradox is reverted when C is weakly Schur-concave. So this property holds for many copulas. However, as stated above, it is not always true (which is also a paradox). For other system structures it is not easy to prove if $\bar{Q}$ is weakly Schur-concave/convex.

We can try to extend the Parrondo paradox to other (stronger) orders by using the comparison results obtained from distortions. For example, in the case of series systems with two independent components, to extend it to the HR order we haver to study the monotonicity of the ratio

$$\frac{\bar{Q}_S(u_1, u_2)}{\bar{Q}_T(u_1, u_2)} = \frac{(u_1 + u_1)^2/4}{u_1 u_2} = \frac{1}{2} + \frac{u_1}{4u_2} + \frac{u_2}{4u_1}.$$

It is easy to see that it is not monotone in $[0, 1]^2$. So the Parrondo paradox cannot be extended to the HR order as can be seen in Fig. 3.16. Note that $T \leq_{HR} S$ holds for two exponential distributions with mean 5 and 1(left) but that it does not hold when the first exponential is replaced with a Weibull (right) with hazard rate $h_1(t) = 4t^3$ for $t \geq 0$. Also note that, in both cases, the limiting value of h_S coincides with the one of the hazard rate of the series system obtained with the best components when $t \to \infty$. This is a well known property in mixture models where the leading term is determined by the best components since the worse components fail before (see Navarro and Hernández 2008a, and the references therein).

Let us come back now to the question of the green line in Fig. 3.14. To answer this question let us consider more general systems with randomized components. They were studied in Navarro et al. (2015). If we have two type of components with reliability function $\bar{F}_X$ and $\bar{F}_Y$ we can consider the deterministic system T_k which have k components from X and $n - k$ from Y. Its reliability function is

$$\bar{F}_{T_k}(t) = \bar{Q}(\underbrace{\bar{F}_X(t), \ldots, \bar{F}_X(t)}_{k \ times}, \underbrace{\bar{F}_Y(t), \ldots, \bar{F}_Y(t)}_{n-k \ times})$$

for $k = 0, \ldots, n$. Here $k = 0$ means that we only use units from Y and $k = n$ that we just use units from X.

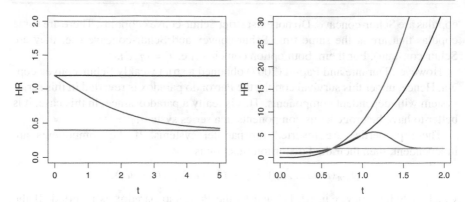

Fig. 3.16 Hazard rate functions for the series systems T (black) and S (blue) in Parrondo paradox for exponential (left) and Weibull (right) distributions. The red and orange plots correspond to the series systems obtained with just units of type 1 (red) or 2 (orange)

Then we can consider the randomized (mixed) system T_K which choose T_k when the random variable $K = k$, where K is a discrete random variable over the set $\{0, \ldots, n\}$.

Note that the systems T and S in the Parrondo paradox when $n = 2$ are obtained as $T = T_1$ (i.e. we choose one component of each type) and $S = T_K$ where $K = 0, 1, 2$ with probabilities $0.25, 0.5, 0.25$, respectively. Note that $E(K) = 1$. Also note that T_1 can be obtained with the atom random variable K which takes the value 1 for sure.

Also note that the red and orange lines in Fig. 3.14 correspond to T_2 and T_0, respectively. As mentioned above, in some cases, these options are unrealistic because they only use units of one type. To have "fair" comparisons, we should impose $E(K) = 1$, that is, we use a 50% of units from each type. Under this condition, which one is the best option for K? The answer is given in the following result extracted from Navarro et al. (2015). There we use the convex (CX), increasing convex (ICX) and increasing concave (ICV) orders. Their definitions and main properties can be seen in Shaked and Shanthikumar (2007).

Proposition 3.18 *If the number k of components of type X is chosen randomly according to the random variables K_1 or K_2 and*

$$\varphi(k) = \bar{Q}(\underbrace{u, \ldots, u}_{k \; times}, \underbrace{v, \ldots, v}_{n-k \; times})$$

is convex (concave) in $\{0, 1, \ldots, n\}$ for all $u, v \in (0, 1)$, then:

(i) $K_1 \leq_{CX} K_2$ implies $T_{K_1} \leq_{ST} T_{K_2}$ ($\geq_{ST}$);
(ii) $X \geq_{ST} Y$ and $K_1 \leq_{ICX} K_2$ ($\leq_{ICV}$) imply $T_{K_1} \leq_{ST} T_{K_2}$ ($\geq_{ST}$).

This result says that if φ is convex, then the more convex K, the better. If the units from X are ST better than the ones from Y, then the convex ordering can be relaxed to the weakly ICX order. These properties are reverted when φ is concave.

In our series system with two independent components

$$\varphi(k) = u^k v^{n-k}$$

which is a convex function of k for any $u, v \in (0, 1)$ since

$$\varphi'(k) = \varphi(k) \log(u/v)$$

and

$$\varphi''(k) = \varphi(k)(\log(u/v))^2 = u^k v^{n-k}(\log(u/v))^2 \geq 0.$$

Hence, from (i) in the preceding proposition, when two randomized options are ordered in the convex order, the respective systems are ST ordered in the same sense. As mentioned above, the first option T is obtained with K_1 which takes the value 1 with probability 1 and the second S with K_2 with probabilities $0.25, 0.5, 0.25$ for $k = 0, 1, 2$, respectively. Another reasonable assumption could be a uniform distribution, that is, $K_3 = 0, 1, 2$ with probability $1/3$. Finally, we could also consider K_4 with probabilities $0.5, 0, 0.5$ for $k = 0, 1, 2$, respectively. In all these options we have $E(K_i) = 1$ for $i = 1, 2, 3, 4$ (i.e. they use the same number of components of each type). It can also be proved (e.g. by plotting their respective probability mass functions) that

$$K_1 \leq_{CX} K_2 \leq_{CX} K_3 \leq_{CX} K_4.$$

Therefore, from (i),

$$T_{K_1} \leq_{ST} T_{K_2} \leq_{ST} T_{K_3} \leq_{ST} T_{K_4}$$

for all $\bar{F}_X, \bar{F}_Y$. Actually, K_4 is the more convex option for K such that $E(K) = 1$. Hence it is always the best option in this system. It corresponds to the green line in Fig. 3.14 and it assumes that the series systems are built with two units of type X or with two units of type Y, randomly. This is the best option for our system and any $\bar{F}_X, \bar{F}_Y$, that is, the green line will be always above the other lines (reliabilities). Note that it is also a "random option" (so it can also be seen as a Parrondo paradox) and that we do not need $\bar{F}_X \geq \bar{F}_Y$. However, if this property holds, this best option could be unreasonable in practice since the 50% of the customers will have a very good system with two good units but the others will have a very bad system built with two bad units. In this case, what should be done in practice? The answer is not easy. Although this best option is very dispersed (we have very good and very bad systems), note that this what we do at home, for example, with the remote control when we have good and bad batteries (we put together the units of the same type). This is the best option for series systems with independent and heterogeneous components.

Problems

1. Compare two systems with IID components by using their signatures. Plot the respective functions to confirm (or reject) the comparisons obtained.
2. Check that an arrow in Fig. 3.2 is correct. Plot the respective functions to confirm (or reject) the comparison obtained.
3. Check that a no-arrow in Fig. 3.2 is correct. Plot the respective functions to confirm (or reject) the comparison obtained.

4. Check that an arrow in Fig. 3.3 is correct. Plot the respective functions to confirm (or reject) the comparison obtained.
5. Check that a no-arrow in Fig. 3.3 is correct. Plot the respective functions to confirm (or reject) the comparison obtained.
6. Check that an arrow in Fig. 3.4 is correct. Plot the respective functions to confirm (or reject) the comparison obtained.
7. Check that a no-arrow in Fig. 3.4 is correct. Plot the respective functions to confirm (or reject) the comparison obtained.
8. Study the orderings for series and parallel systems with ID components for a given bivariate survival copula. Plot the respective functions to confirm (or reject) the comparisons obtained.
9. Study the orderings for series systems with ID components and two different bivariate survival copulas. Plot the respective functions to confirm (or reject) the comparisons obtained.
10. Study the orderings for two systems with ID components for a given trivariate survival copula. Plot the respective functions to confirm (or reject) the comparisons obtained.
11. Study the effect of the dependence parameter of a copula in the reliability of a system with ID components. Plot the respective functions to confirm (or reject) the comparisons obtained.
12. Study the orderings $X_{1:3} \leq_{HR} X_{2:3} \leq_{HR} X_{3:3}$ for ID components and a survival copula $\widehat{C}$.
13. Find an EXC copula for which $X_{1:2} \leq_{HR} X_{2:2}$ does not hold in the ID case. Plot the hazard rate functions to confirm that this comparison does not hold.
14. Check that a number in Table 3.4 is correct. Plot the respective functions to confirm (or reject) the comparison obtained.
15. Check that a number in Table 3.5 is correct. Plot the respective functions to confirm (or reject) the comparisons obtained.
16. Check that an arrow in Fig. 3.11 is correct. Plot the respective functions to confirm (or reject) the comparison obtained.
17. Check that a no-arrow in Fig. 3.11 is correct. Plot the respective functions to confirm (or reject) the comparison obtained.
18. Compare $X_{1:2}$, X_1, X_2 and $X_{2:2}$ for a fixed bivariate survival copula $\widehat{C}$ and arbitrary distributions F_1, F_2.
19. Compare two semi-coherent systems of order 3 for a fixed trivariate survival copula $\widehat{C}$ and arbitrary distributions F_1, F_2, F_3.
20. Compare two systems by using RR-plots.
21. Confirm the Parrondo paradox in series systems with independent components.
22. Confirm the Parrondo paradox in series systems with dependent components and an Archimedean copula.
23. Study the Parrondo paradox in a non-series system with independent components. Plot the respective reliability functions.
24. Prove that the Parrondo paradox holds for series systems with n independent components.
25. Prove that the Parrondo paradox is reverted for parallel systems with n independent components.

Aging Properties

4

Abstract

In this chapter we study the process of growing old for the system and the components. To this end we use the main aging functions and the associated aging classes (IFR, NBU, DMRL, ILR and their respective dual classes). In particular we state conditions for the preservation of some of these aging classes under the formation of coherent systems. We also consider different system residual and inactivity times. The limiting behavior (when the time increases) of some system aging functions are studied as well. The same technique can be used to get bounds for them.

4.1 Main Aging Classes

First we give the definitions and the main properties of the aging classes considered here. Note that they can be used to describe the aging process of the system and the components. For more properties and applications we refer the readers to Belzunce et al. (2016), Müller and Stoyan (2002) and Shaked and Shanthikumar (2007).

Let X be a non-negative random variable (r.v.) representing the lifetime of a unit or a system (some aging classes can also be defined for r.v. that can take negative values). Let F be its distribution function and let $\bar{F} = 1 - F$ be its reliability function. As in the first chapter, we consider its residual lifetime $X_t = (X - t \mid X > t)$ and the MRL function $m(t) = E(X - t \mid X > t)$ for $t \geq 0$ (when these expectations exist). If F is absolutely continuous, then $f = F'$ represents its pdf and $h = f/\bar{F}$ its hazard (or failure) rate function. Then we consider the following aging classes.

Definition 4.1 X is said to be **Increasing (Decreasing) Failure Rate**, IFR (DFR), if $X_s \geq_{ST} X_t$ ($\leq_{ST}$) for all $s \leq t$ (such that these conditional r.v. exist).

By historical reasons, the researchers prefer to use "failure rate" for this class but "hazard rate" for the order. However, if you prefer, these classes can also be written as IHR/DHR. The HR order could also be written as the FR order. Note that IFR means that the used units with residual lifetimes X_t are ST decreasing in t. These kind of conditions represent positive (or natural) aging properties where the unit get worse when the time goes on. Here "positive" does not mean "good". The reverse conditions represent "negative" (or unnatural) aging behaviors. The IFR condition can also be written as

$$\bar{F}(x + s)\bar{F}(t) - \bar{F}(x + t)\bar{F}(s) \geq 0$$

for all $s \leq t$ and all $x \geq 0$. This equality is reversed for the DFR class. The IFR (DFR) is also characterized by a log-concave (log-convex) reliability, that is, $\ln F$ is concave (convex). If F is abs. cont., then IFR (DFR) is characterized by an increasing (decreasing) hazard (or failure) rate. Obviously, this property explains its name. The exponential distribution belongs to both classes (since it has a constant hazard rate in $[0, \infty)$). Moreover, the ST order used in the definition can be replaced with the HR order since $h_{X_t}(x) = h(x + t)$ for all $t, x \geq 0$.

As we will see later, these classes are not preserved under the formation of coherent systems. So we could consider the following weaker conditions.

Definition 4.2 X is said to be **New Better (Worse) than Used**, NBU (NWU), if $X \geq_{ST} X_t$ ($\leq_{ST}$) for all $t \geq 0$ (such that these conditional r.v. exist).

Note that in these classes we need to assume that X is non-negative. Again, the exponential distribution belongs to both classes since $X =_{ST} X_t$ for all $t \geq 0$. The condition for the NBU property can also be written as

$$\bar{F}(x)\bar{F}(t) \geq \bar{F}(x + t),$$

that is, $\ln \bar{F}$ is superaditive. Obviously, IFR implies NBU and DFR implies NWU but the reverse implications do not hold. Between these classes we have the ones defined as follows.

Definition 4.3 X is said to be **Increasing (Decreasing) Failure Rate Average, IFRA (DFRA)**, if the function

$$A(t) = \frac{1}{t} \int_0^t h(x)dx = -\frac{1}{t} \ln \bar{F}(t)$$

is increasing (decreasing) for all $t \geq 0$.

It can be proved that these conditions are equivalent to

$$\bar{F}(ct) \geq \bar{F}^c(t) \quad (\leq) \tag{4.1}$$

for all $c \in (0, 1)$ and all $t \geq 0$. As in the preceding case, the exponential distribution belongs to both classes and we have

$$IFR \Rightarrow IFRA \Rightarrow NBU$$

and
$$DFR \Rightarrow DFRA \Rightarrow NWU.$$

The classes based on the reversed failure rate are defined in a similar way. For example, X is DRFR (IRFR) is $\bar{h} = f/F$ is decreasing (increasing). Block et al. (1998) proved that there are no non-negative random variables which have the IRFR property.

For the mean residual life function we have the following classes.

Definition 4.4 X is said to be **Increasing (Decreasing) Mean Residual Life**, IMRL (DMRL), if the MRL function

$$m(t) = E(X - t|X > t) = \frac{1}{\bar{F}(t)} \int_0^t \bar{F}(x)dx$$

is increasing (decreasing) for all $t \geq 0$.

For the exponential distribution we have $m(t) = E(X - t|X > t) = E(X)$ and so it belongs to both classes. Note that here the positive (negative) aging is represented by the DMRL (IMRL) class. Weaker classes can be obtained as follows.

Definition 4.5 X is said to be **New Better (Worse) than Used in Expectations,** NBUE (NWUE) if $E(X) \geq E(X - t|X > t)$ $(\leq)$ for all $t \geq 0$.

Clearly, we have
$$IFR \Rightarrow DMRL \Rightarrow NBUE$$

and

$$DFR \Rightarrow IMRL \Rightarrow NWUE.$$

The last classes are the strongest ones and are related with the LR order.

Definition 4.6 X is said to be **Increasing (Decreasing) Likelihood Ratio**, ILR (DLR), if $X_s \geq_{LR} X_t$ $(\leq_{LR})$ for all $0 \leq s \leq t$ (such that these conditional r.v. exist).

Clearly, the condition for the ILR (DLR) class implies the IFR (DFR) property (since the LR order is stronger than the ST order) and is equivalent to: f is log-concave (log-convex). They can also be stated in terms of the Glaser's eta function $\eta(t) = -f'(t)/f(t)$, ILR means that η is increasing and DLR that it is decreasing.

The relationships between these classes are summarized in Table 4.1. The dual classes satisfy similar relationships.

For more properties on aging classes see Bryson and Siddiqui (1969), Barlow and Proschan (1975), Block et al. (2006) and Navarro et al. (2008). In many applications, the units (or the systems) do not have monotone failure rate (or mean residual life) functions. Actually, very often, they have a **Bathtub shaped Failure Rate** (BFR), that is, $h(t)$ is decreasing for $t \in [0, t_1]$, is constant in $[t_1, t_2]$, and is increasing in $[t_2, \infty)$ for some $t_1 \leq t_2$. However, few distributions have this shape for their hazard

Table 4.1 Relationships among the main positive aging classes

ILR	$\Rightarrow$	IFR	$\Rightarrow$	IFRA	$\Rightarrow$	NBU
$\Downarrow$		$\Downarrow$				$\Downarrow$
DRFR		DMRL		$\Rightarrow$		NBUE

rate functions. Some authors try to solve this problem by adding parameters to well known distributions. Others prefer to consider mixtures (or generalized mixtures) of populations which also explain why this shape appears (when the populations contains different units), see e.g. Navarro and Hernández (2004, 2008a, b) and the references therein. We will also see that this shape appears when we consider systems.

4.2 Systems with ID Components

First of all, we recall that if T is the lifetime of a system with ID component lifetimes having a common distribution function F, then, from Theorem 2.11, the system distribution function can be written as $F_T(t) = q(F(t))$ for all t and for a continuous and increasing distortion function $q : [0, 1] \rightarrow [0, 1]$ such that $q(0) = 0$ and $q(1) = 1$. The respective reliability functions satisfy $\bar{F}_T(t) = \bar{q}(\bar{F}(t))$ for all t, where $\bar{q}(u) := 1 - q(1 - u)$ for all $u \in [0, 1]$ is another distortion function.

Then we can use the preservation results for distorted distributions obtained in Navarro et al. (2014) to study the preservation of aging classes in systems. They can be stated as follows. We say that an aging class $\mathcal{A}$ is **preserved** by a distortion q iff $q(F) \in \mathcal{A}$ for all $F \in \mathcal{A}$.

Theorem 4.1 *Let* $F_q = q(F)$ *be a distorted distribution. Then:*

(i) *The IFR (DFR) class is preserved by* q *iff* $\alpha(u) = u\bar{q}'(u)/\bar{q}(u)$ *is decreasing (increasing) for* $u \in (0, 1)$.

(ii) *The DRFR class is preserved by* q *iff* $\bar{\alpha}(u) = uq'(u)/q(u)$ *is decreasing in* $(0, 1)$.

(iii) *The NBU (NWU) class is preserved by* q *iff* $\bar{q}$ *is submultiplicative (supermultiplicative), that is,*

$$\bar{q}(uv) \leq \bar{q}(u)\bar{q}(v), \ (\geq) \text{ for all } u, v \in [0, 1]. \tag{4.2}$$

(iv) *The IFRA (DFRA) class is preserved by* q *iff* $\bar{q}$ *satisfies*

$$\bar{q}(u^c) \geq (\bar{q}(u))^c, \ (\leq) \text{ for all } u, c \in [0, 1]. \tag{4.3}$$

(v) *If* F *is absolutely continuous and ILR and there exists* $u_0 \in [0, 1]$ *such that* $\beta(u) = u\bar{q}''(u)/\bar{q}'(u)$ *is non-negative and decreasing in* $[0, u_0]$ *and* $\bar{\beta}(u) = (1 - u)\bar{q}''(u)/\bar{q}'(u)$ *is non-positive and decreasing in* $[u_0, 1]$, *then* F_q *is ILR.*

Proof To prove (i), we use (2.34) which allows us to write the respective hazard rate functions as

$$h_q(t) = \alpha(\bar{F}(t))h(t).$$

Then, if we assume that α is decreasing (increasing) and h is increasing, then h_q is increasing (decreasing) since $\bar{F}$ is always decreasing.

Conversely, if we assume that the IFR (DFR) class is preserved, then it is preserved for a standard exponential distribution which has reliability function $\bar{F}(t) = \exp(-t)$ and hazard rate function $h(t) = 1$ for $t \geq 0$. Hence $h_q(t) = \alpha(\exp(-t))$ is increasing (decreasing) for $t \in (0, \infty)$ and so α decreases (increases) in $(0, 1)$.

The proof of (ii) is analogous to that of (i).

To prove (iii), we note that F_q is NBU (NWU) iff

$$\bar{q}(\bar{F}(x + t)) \leq \bar{q}(\bar{F}(x))\bar{q}(\bar{F}(t)), \ (\geq) \text{ for all } x, t \geq 0.$$

If we assume that F is NBU and $\bar{q}$ is submultiplicative, that is, it satisfies (4.2), then

$$\bar{q}(\bar{F}(x + t)) \leq \bar{q}(\bar{F}(x)\bar{F}(t)) \leq \bar{q}(\bar{F}(x))\bar{q}(\bar{F}(t))$$

for all $x, t \geq 0$ and so F_q is NBU.

Conversely, if the NBU class is preserved, then it is preserved for a standard exponential distribution, that is,

$$\bar{q}(\bar{F}(x + t)) \leq \bar{q}(\bar{F}(x))\bar{q}(\bar{F}(t))$$

holds for all $t, x \geq 0$ and $\bar{F}(z) = \exp(-z)$. Hence

$$\bar{q}(e^{-x}e^{-t}) \leq \bar{q}(e^{-x})\bar{q}(e^{-t})$$

holds for all $t, x \geq 0$ and so (4.2) holds. The proof for the NWU class is similar.

Finally, to prove (iv) we note that, from (4.1), F_q is IFRA (DFRA) iff

$$\bar{q}(\bar{F}(ct)) \geq (\bar{q}(\bar{F}(t)))^c$$

for all $c \in (0, 1)$ and all $t \geq 0$.

If we assume that F is IFRA and (4.3) holds, then

$$\bar{q}(\bar{F}(ct)) \geq \bar{q}(\bar{F}^c(t)) \geq (\bar{q}(\bar{F}(t)))^c$$

for all $c \in (0, 1)$ and all $t \geq 0$.

Conversely, if the IFRA class is preserved, so is it for the standard exponential distribution. Thus we get

$$\bar{q}(e^{-ct}) \geq (\bar{q}(e^{-t}))^c$$

for all $c \in (0, 1)$ and all $t \geq 0$. Hence, (4.3) holds. The proof for the DFRA class is analogous.

The proof of (v) can be seen in Navarro et al. (2014). $\qquad\square$

Note that we can also obtain reverse results. For example, if the distorted distribution F_q is IFR and α is increasing, then F is also IFR. Also note that if both IFR and DFR classes are preserved, then the function α is constant in $(0, 1)$ and so $\bar{q}(u) = u^c$ holds for $u \in [0, 1]$ and $c > 0$ (PHR model). This property is satisfied by series systems with IID components and so both classes are preserved for them. However, note that if α is increasing (or decreasing) but it is not constant, then just the DFR (IFR) class is preserved. If α is not monotone, then neither of them are preserved.

Remark 4.1 We can prove that if the IFR class is preserved, that is, α is decreasing, then the NBU class is also preserved. Note that if the IFR is preserved, as the standard exponential distribution is IFR, we have that $\bar{q}(e^{-t})$ is IFR. Then, it is also NBU and so

$$\bar{q}(e^{-(t+x)}) = \bar{q}(e^{-t}e^{-x}) \leq \bar{q}(e^{-t})\bar{q}(e^{-x})$$

holds for all $x, t > 0$. Hence we get (4.2). A similar property holds for the NWU class (it is preserved when the DFR class is preserved).

Analogously, we can prove that if the IFR class is preserved, then the IFRA class is preserved too. Note that if the IFR is preserved, as the exponential distribution is IFR, we have that $\bar{q}(e^{-ct})$ is IFR for all $c > 0$. Then, it is also IFRA and so

$$\bar{q}(e^{-ct}) = \bar{q}((e^{-t})^c) \geq (\bar{q}(e^{-t}))^c$$

holds for all $c \in (0, 1)$ and $t > 0$. Hence we get (4.3). A similar property holds for the DFRA class (it is preserved when the DFR class is preserved).

Let us see some examples. In the first one we see that the IFR class is preserved under the formation of k-out-of-n systems with IID components. This property was proved by Esary and Proschan (1963). As a consequence, the DFR class is not preserved, except in the case of series systems.

Example 4.1 Let us consider the k-out-of-n system with IID$\sim F$ components and lifetime $X_{i:n}$ (i.e. $k = n - i + 1$). From (2.8) and (2.12), its hazard rate function is

$$h_{i:n}(t) = \frac{i\binom{n}{i}f(t)F^{i-1}(t)\bar{F}^{n-i}(t)}{\sum_{j=0}^{i-1}\binom{n}{j}F^j(t)\bar{F}^{n-j}(t)} = \frac{i\binom{n}{i}F^{i-1}(t)\bar{F}^{n-i+1}(t)}{\sum_{j=0}^{i-1}\binom{n}{j}F^j(t)\bar{F}^{n-j}(t)}\frac{f(t)}{\bar{F}(t)} = \alpha(\bar{F}(t))h(t),$$

where h is the common hazard rate of the components and

$$\alpha(u) = \frac{i\binom{n}{i}}{\sum_{j=0}^{i-1}\binom{n}{j}(1-u)^{j-i+1}u^{i-j-1}} = \frac{i\binom{n}{i}}{\sum_{j=0}^{i-1}\binom{n}{j}w^{i-j-1}(u)}$$

being $w(u) := u/(1-u)$ an increasing function in $(0, 1)$. Hence, α is decreasing in $(0, 1)$ and so the IFR class is always preserved. As a consequence, the NBU and IFRA classes are preserved as well and the DFR class is not preserved except in the case $i = 1$ (series systems) where $\alpha(u) = n$. Note that from the results given in the preceding chapter, the HR order is also preserved under the formation of k-out-of-n systems with IID components. It can be proved that DRFR, IRFR, NWU and DFRA classes are not preserved.

To confirm these results we consider k-out-of-3 systems for $k = 1, 2, 3$, with IID components having a common standard exponential distribution and constant hazard rate. In Fig. 4.1, we plot their respective α functions (left) and hazard rate functions (right). Note that the components are DFR but that $X_{2:3}$ and $X_{3:3}$ are not DFR. Also note that the limiting behavior of $h_{3:3}$ (black line, right) coincides with that of h (dotted line, right). The R code to plot these functions is the following. It can be changed to get similar plots for other systems/distributions.

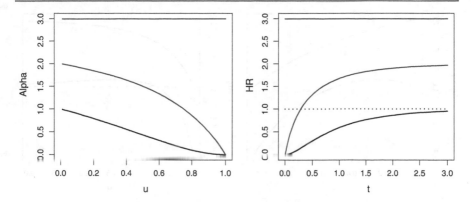

Fig. 4.1 Alpha functions (left) and hazard rate functions (right) of k-out-of-3 systems for $k = 1, 2, 3$ (black, blue, red) with IID components having a common standard exponential distribution in Example 4.1. The dotted line represents the hazard rate of the components

```
# Alpha functions:
R<-function(t) exp(-t)
q13<-function(u) u^3
q23<-function(u) 3*u^2-2*u^3
q33<-function(u) 3*u-3*u^2+u^3
q13p<-function(u) 3*u^2
q23p<-function(u) 6*u-6*u^2
q33p<-function(u) 3-6*u+3*u^2
a13<-function(u) u*q13p(u)/q13(u)
a23<-function(u) u*q23p(u)/q23(u)
a33<-function(u) u*q33p(u)/q33(u)
curve(a33(x),xlab='u',ylab='Alpha',ylim=c(0,3),lwd=2)
curve(a23(x),add=T,col='blue',lwd=2)
curve(a13(x),add=T,col='red',lwd=2)

#Hazard rate functions:
f<-function(t) exp(-t)
h<-function(t) f(t)/R(t)
curve(a33(R(x))*h(x),ylim=c(0,3),ylab='HR',xlab='t',0,3,lwd=2)
curve(a23(R(x))*h(x),add=T,col='blue',lwd=2)
curve(a13(R(x))*h(x),add=T,col='red',lwd=2)
curve(h(x),add=T,lty=3,lwd=2)
```

In the following example we show that the IFR class can also be preserved in other systems but that we can also find systems where this class is not preserved.

Example 4.2 Let us consider the systems with IID$\sim F$ components and with lifetimes $T_1 = \min(X_1, \max(X_2, X_3))$ and $T_2 = \max(X_1, \min(X_2, X_3))$. Their

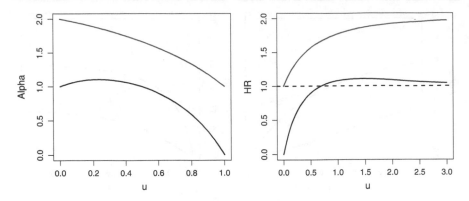

Fig. 4.2 Alpha functions (left) and hazard rate functions (right) for the systems T_1 (blue) and T_2 (black) with IID components having a common standard exponential distribution in Example 4.2. The dotted line represents the common hazard rate of the components

respective distortion functions are $\bar{q}_1(u) = 2u^2 - u^3$ and $\bar{q}_2(u) = u + u^2 - u^3$ and their alpha-functions

$$\alpha_1(u) = \frac{u\bar{q}_1'(u)}{\bar{q}_1(u)} = \frac{4 - 3u}{2 - u}$$

and

$$\alpha_2(u) = \frac{u\bar{q}_2'(u)}{\bar{q}_2(u)} = \frac{1 + 2u - 3u^2}{1 + u - u^2}$$

for $u \in (0, 1)$. It can be proved that α_1 is strictly decreasing but that α_2 is not monotone in $(0, 1)$ (see Fig. 4.2, left). Therefore, the IFR class is preserved in T_1 but it is not preserved in T_2. Moreover, the DFR class is not preserved in these systems.

This is a very surprising property since when we put (independent) components with "natural" aging (IFR) into system T_2, then we get a system that does not have this natural aging property. Actually, as we can see in Fig. 4.2, right, the system T_2 improves when $t > t_0 \approx 1.444$. The meaning of this fact can be seen in Fig. 4.3, left, where we plot the reliability functions of the residual lifetimes $(T_2)_t = (T_2 - t | T > t)$ for $t = 0$ (black line, new units), $t = 0.2$ (blue line), $t = 1.444$ (green line, more reliable age), and $t = 5$ (red line, liming behavior). Note that T_2 seems to be NBU since its residual lifetimes are ST worse than T_2. We can confirm this property by using Theorem 4.1 showing that $\bar{q}_2$ is submultiplicative. We will see in the next section that this property holds for all the coherent systems with independent components.

The hazard rate functions for T_2 are plotted in Fig. 4.3, right, when the components have Weibull distributions $F(t) = 1 - \exp(-t^\beta)$, $t \geq 0$, with shape parameter $\beta = 0.5, 1, 1.2, 2$ (green, black, red, blue). The dotted lines represent the hazard rate of the components. Note that the IFR and DFR classes are sometimes preserved but that this is not always the case. Also note that the limiting behavior of the hazard rate functions of the systems and the components coincide. We will show later that this is a general property for this system.

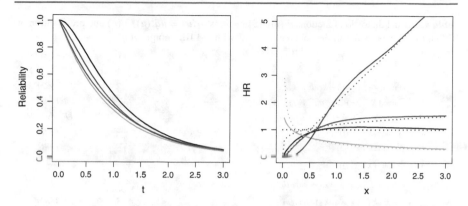

Fig. 4.3 Reliability functions (left) for the system T_2 with IID components having a common standard exponential distribution in Example 4.2 when new (black line) and with ages $t = 0.2, 1.444, 5$ (blue, green, red). Hazard rate functions (right) for T_2 when the components have Weibull distributions with shape parameter $\beta = 0.5, 1, 1.2, 2$ (green, black, red, blue). The dotted lines represent the hazard rate functions of the components

The code to get these plots is the following:

```
#Reliability functions (Figure 4.3, left):
R<-function(t) exp(-t)
q2<-function(u) u+u^2-u^ 3
f<-function(t) exp(-t)
h<-function(t) f(t)/R(t)
R2<-function(t) q2(R(t))
curve(R2(x),xlab='t',ylab='Reliability',ylim=c(0,1),0,3,lwd=2)
curve(R2(x+0.2)/R2(0.2),col='blue',add=T,lwd=2)
curve(R2(x+1.444)/R2(1.444),col='green',add=T,lwd=2)
curve(R2(x+5)/R2(5),col='red',add=T,lwd=2)


# Hazard rate functions (Figure 4.3, right):
R<-function(t,b) exp(-t^b)
h<-function(t,b) b*t^(b-1)
curve(a2(R(x),1)*h(x,1),ylab='HR',ylim=c(0,5),0,3,lwd=2)
curve(h(x,1),add=T,lty=3,lwd=2)
curve(a2(R(x,1.2))*h(x,1.2),add=T,col='red',lwd=2)
curve(h(x,1.2),add=T,lty=3,col='red',lwd=2)
curve(a2(R(x,0.5))*h(x,0.5),add=T,col='green',lwd=2)
curve(h(x,0.5),add=T,lty=3,col='green',lwd=2)
curve(a2(R(x,2))*h(x,2),add=T,col='blue',lwd=2)
curve(h(x,2),add=T,lty=3,col='blue',lwd=2)
```

By using the tools showed in the preceding example we can study the preservation of IFR and DFR classes in systems with IID components. The results for systems

Table 4.2 Dual distortion functions $\bar{q}_i$, alpha functions $\alpha_i(u) = u\bar{q}_i'(u)/\bar{q}_i(u)$ and preservation of IFR and DFR classes for all the coherent systems with 1-4 IID components

i	T_i	$\bar{q}_i$	α_i	*Preserved*
1	$X_{1:1} = X_1$	u	1	$IFR\&DFR$
2	$X_{1:2} = \min(X_1, X_2)$ $(2 - series)$	u^2	2	$IFR\&DFR$
3	$X_{2:2} = \max(X_1, X_2)$ $(2 - parallel)$	$2u - u^2$	$\frac{2-2u}{2-u}$	IFR
4	$X_{1:3} = \min(X_1, X_2, X_3)$ $(3 - series)$	u^3	3	$IFR\&DFR$
5	$\min(X_1, \max(X_2, X_3))$	$2u^2 - u^3$	$\frac{4-3u}{2-u}$	IFR
6	$X_{2:3}(2 - out - of - 3)$	$3u^2 - 2u^3$	$\frac{6-6u}{3-2u}$	IFR
7	$\max(X_1, \min(X_2, X_3))$	$u + u^2 - u^3$	$\frac{1+2u-3u^2}{1+u-u^2}$	–
8	$X_{3:3} = \max(X_1, X_2, X_3)$ $(3 - parallel)$	$3u - 3u^2 + u^3$	$\frac{3-6u+3u^2}{3-3u+u^2}$	IFR
9	$X_{1:4} = \min(X_1, X_2, X_3, X_4)$ $(series)$	u^4	4	$IFR\&DFR$
10	$\max(\min(X_1, X_2, X_3), \min(X_2, X_3, X_4))$	$2u^3 - u^4$	$\frac{6-4u}{2-u}$	IFR
11	$\min(X_{2:3}, X_4)$	$3u^3 - 2u^4$	$\frac{9-8u}{3-2u}$	IFR
12	$\min(X_1, \max(X_2, X_3), \max(X_3, X_4))$	$u^2 + u^3 - u^4$	$\frac{2+3u-4u^2}{1+u-u^2}$	–
13	$\min(X_1, \max(X_2, X_3, X_4))$	$3u^2 - 3u^3 + u^4$	$\frac{6-9u+4u^2}{3-3u+u^2}$	IFR
14	$X_{2:4}$ $(3 - out - of - 4)$	$4u^3 - 3u^4$	$\frac{12-12u}{4-3u}$	IFR
15	$\max(\min(X_1, X_2), \min(X_1, X_3, X_4),$ $\min(X_2, X_3, X_4))$	$u^2 + 2u^3 - 2u^4$	$\frac{2+6u-8u^2}{1+2u-2u^2}$	–
16	$\max(\min(X_1, X_2), \min(X_3, X_4))$	$2u^2 - u^4$	$\frac{4-4u^2}{2-u^2}$	IFR
17	$\max(\min(X_1, X_2), \min(X_1, X_3),$ $\min(X_2, X_3, X_4))$	$2u^2 - u^4$	$\frac{4-4u^2}{2-u^2}$	IFR
18	$\max(\min(X_1, X_2), \min(X_2, X_3),$ $\min(X_3, X_4))$	$3u^2 - 2u^3$	$\frac{6-6u}{3-2u}$	IFR
19	$\max(\min(X_1, \max(X_2, X_3, X_4)),$ $\min(X_2, X_3, X_4))$	$3u^2 - 2u^3$	$\frac{6-6u}{3-2u}$	IFR
20	$\min(\max(X_1, X_2), \max(X_1, X_3),$ $\max(X_2, X_3, X_4))$	$4u^2 - 4u^3 + u^4$	$\frac{8-12u+4u^2}{4-4u+u^2}$	IFR
21	$\min(\max(X_1, X_2), \max(X_3, X_4))$	$4u^2 - 4u^3 + u^4$	$\frac{8-12u+4u^2}{4-4u+u^2}$	IFR
22	$\min(\max(X_1, X_2), \max(X_1, X_3, X_4),$ $\max(X_2, X_3, X_4))$	$5u^2 - 6u^3 + 2u^4$	$\frac{10-18u+8u^2}{5-6u+2u^2}$	IFR
23	$X_{3:4}(2 - out - of - 4)$	$6u^2 - 8u^3 + 3u^4$	$\frac{12-24u+12u^2}{6-8u+3u^2}$	IFR
24	$\max(X_1, \min(X_2, X_3, X_4))$	$u + u^3 - u^4$	$\frac{1+3u^2-4u^3}{1+u^2-u^3}$	–
25	$\max(X_1, \min(X_2, X_3), \min(X_3, X_4))$	$u + 2u^2 - 3u^3 + u^4$	$\frac{1+4u-9u^2+4u^3}{1+2u-3u^2+u^3}$	–
26	$\max(X_{2:3}, X_4)$	$u + 3u^2 - 5u^3 + 2u^4$	$\frac{1+6u-15u^2+8u^3}{1+3u-5u^2+2u^3}$	–
27	$\min(\max(X_1, X_2, X_3), \max(X_2, X_3, X_4))$	$2u - 2u^3 + u^4$	$\frac{2-6u^2+4u^3}{2-2u^2+u^3}$	IFR
28	$X_{4:4} = \max(X_1, X_2, X_3, X_4)(4 - parallel)$	$4u - 6u^2 + 4u^3 - u^4$	$\frac{4-12u+12u^2-4u^3}{4-6u+4u^2-u^3}$	IFR

with 1-4 components are in Table 4.2. Recall that, in this case, the NBU class is always preserved, that the IFR is preserved in k-out-of-n systems, and that the DFR class is preserved as well in series systems. Conditions for the preservation of the NBUE and NWUE classes were obtained in Lindqvist and Samaniego (2019). There it is also proved that the DFR class is not preserved by any coherent systems with IID components other than series systems.

In the last example we show that we can also study systems with dependent ID components. This example proves that the IFR class is not preserved in k-out-of-n systems with dependent components and that the DFR class can be preserved in these systems.

Example 4.3 We consider now series and parallel systems with two dependent components having standard exponential distributions and the Clayton–Oakes survival copula in (3.12). Their respective dual distortion functions are

$$\bar{q}_{1:2}(u) = \frac{u}{2-u}$$

and

$$\bar{q}_{2:2}(u) = 2u - \frac{u}{2-u}$$

for $u \in [0, 1]$. Then their respective alpha functions are

$$\alpha_{1:2}(u) = \frac{2}{2-u}$$

and

$$\alpha_{2:2}(u) = \frac{2u^2 - 8u + 6}{2u^2 - 7u + 6}.$$

They are plotted in Fig. 4.4, left. We can see there that $\alpha_{1:2}$ is increasing but that $\alpha_{2:2}$ is decreasing. Therefore, the IFR class is preserved in $X_{2:2}$ but it is not preserved in $X_{1:2}$. We have the opposite for the DFR class. The hazard rates for standard exponential components are plotted in Fig. 4.4, right. As the components are both IFR and DFR, we get that $h_{1:2}$ is decreasing and $h_{2:2}$ is increasing. Moreover, they have the same limiting behavior as that of the hazard rate of the components. Also

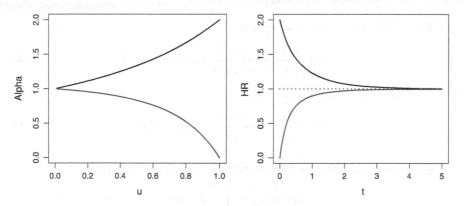

Fig. 4.4 Alpha functions (left) and hazard rate functions for the series (black) and parallel (blue) systems in Example 4.3. The dotted line represents the hazard rate of the components

note that $X_{1:2} \leq_{HR} X_i \leq_{HR} X_{2:2}$ holds for any F (since the alpha functions are ordered). The code to get these plots is the following:

```
# Alpha functions:
C<-function(u,v) u*v/(u+v-u*v)
q12<-function(u) C(u,u)
q22<-function(u) 2*u-C(u,u)
q12p<-function(u) 2/(2-u)^2
q22p<-function(u) 2-q12p(u)
a12<-function(u) u*q12p(u)/q12(u)
a22<-function(u) u*q22p(u)/q22(u)
curve(a12(x),xlab='u',ylab='Alpha',ylim=c(0,2),lwd=2)
curve(a22(x),add=T,col='blue',lwd=2)

# Hazard rate functions
R<-function(t) exp(-t)
f<-function(t) exp(-t)
h<-function(t) f(t)/R(t)
curve(a12(R(x))*h(x),ylab='HR',xlab='t',ylim=c(0,2),0,5,lwd=2)
curve(a22(R(x))*h(x),add=T,col='blue',lwd=2)
curve(h(x),add=T,lty=3,lwd=2)
```

We can also study the preservation of the DMRL/IMRL classes. Abouammoh and El-Neweihi (1986) proved that the DMRL class is preserved under the formation of parallel systems with IID components. This result was extended in Navarro (2018a) given sufficient conditions for the preservation of DMRL/IMRL classes. They can be stated as follows.

Theorem 4.2 *Let $F_q = q(F)$ be a distorted distribution and let $\bar{q}$ be its associated dual distortion. Then:*

(i) The DMRL class is preserved by q if

$$\sup_{u\in(0,v]} \frac{\bar{q}(u)}{u} \leq \frac{\bar{q}^2(v)}{v^2\bar{q}'(v)} \text{ for all } v \in (0,1). \qquad (4.4)$$

(ii) The IMRL class is preserved by q if

$$\inf_{u\in(0,v]} \frac{\bar{q}(u)}{u} \geq \frac{\bar{q}^2(v)}{v^2\bar{q}'(v)} \text{ for all } v \in (0,1). \qquad (4.5)$$

Proof First, we recall that the MRL function of F can we obtained as

$$m(t) = \frac{1}{\bar{F}(t)} \int_t^\infty \bar{F}(x)dx$$

for t such that $\bar{F}(t) > 0$. A similar expression holds for the MRL m_q of F_q. Hence

$$m'(t) =_{sign} -\bar{F}^2(t) + f(t) \int_t^\infty \bar{F}(x) dx.$$

Therefore, F is DMRL iff

$$\frac{f(t)}{\bar{F}^2(t)} \int_t^\infty \bar{F}(x) dx \leq 1 \qquad (4.6)$$

for all $t \geq 0$. We assume that this property holds and then we want to prove that a similar condition holds for F_q, that is,

$$\frac{f(t)\bar{q}'(\bar{F}(t))}{\bar{q}^2(\bar{F}(t))} \int_t^\infty \bar{q}(\bar{F}(x)) dx \leq 1$$

for all $t \geq 0$. We note that

$$\frac{f(t)\bar{q}'(\bar{F}(t))}{\bar{q}^2(\bar{F}(t))} \int_t^\infty \bar{q}(\bar{F}(x)) dx = \frac{f(t)\bar{q}'(\bar{F}(t))}{\bar{q}^2(\bar{F}(t))} \int_t^\infty \frac{\bar{q}(\bar{F}(x))}{\bar{F}(x)} \bar{F}(x) dx$$

and then, from (4.4), we get

$$\frac{f(t)\bar{q}'(\bar{F}(t))}{\bar{q}^2(\bar{F}(t))} \int_t^\infty \bar{q}(\bar{F}(x)) dx \leq \frac{f(t)\bar{q}'(\bar{F}(t))}{\bar{q}^2(\bar{F}(t))} \frac{\bar{q}^2(\bar{F}(t))}{\bar{F}^2(t)\bar{q}'(\bar{F}(t))} \int_t^\infty \bar{F}(x) dx \leq 1$$

where the last inequality holds from (4.6).

The proof of (ii) is similar. $\qquad\square$

The result for parallel systems with IID components given in Abouammoh and El-Neweihi (1986) can be obtained as follows.

Example 4.4 If $X_1, \ldots, X_n$ are IID$\sim F$, then the distortion functions of $X_{n:n} = \max(X_1, \ldots, X_n)$ are $q_{n:n}(u) = u^n$ and

$$\bar{q}_{n:n}(u) = 1 - q_{n:n}(1-u) = 1 - (1-u)^n$$

for $u \in [0, 1]$. To check if (4.4) holds, we consider the function

$$g_n(u) = \frac{\bar{q}_{n:n}(u)}{u} = \frac{1 - (1-u)^n}{u} = \binom{n}{1} - \binom{n}{2}u + \binom{n}{3}u^2 - \cdots + (-1)^{n+1}\binom{n}{n}u^{n-1}.$$

It is a decreasing function in $[0, 1]$ and so

$$\sup_{u \in (0, v]} \frac{\bar{q}_{n:n}(u)}{u} = \lim_{u \to 0^+} g_n(u) = g_n(0) = n.$$

Moreover, a straightforward calculation shows that

$$\frac{\bar{q}_{n:n}^2(v)}{v^2 \bar{q}'_{n:n}(v)} \geq \lim_{v \to 0^+} \frac{\bar{q}_{n:n}^2(v)}{v^2 \bar{q}'_{n:n}(v)} = \frac{g_n^2(0)}{\bar{q}'_{n:n}(0)} = n$$

for $u \in [0, 1]$. Therefore, (4.4) holds and the DMRL class is preserved.

The next example (extracted from Navarro 2018a) shows that the DMRL is also preserved in other systems.

Example 4.5 Let us consider the coherent system with lifetime

$$T = \min(\max(X_1, X_2, X_3), \max(X_2, X_3, X_4))$$

and IID$\sim F$ components. Then $\bar{q}(u) = 2u - 2u^3 + u^4$ (see system 27 in Table 4.2). Hence

$$g_1(u) = \frac{\bar{q}(u)}{u} = 2 - 2u^2 + u^3$$

which is a decreasing function in $(0, 1)$. Therefore

$$\sup_{u \in (0,v]} \frac{\bar{q}(u)}{u} = \lim_{u \to 0^+} g_1(u) = g_1(0) = 2.$$

Moreover,

$$g_2(v) = \frac{\bar{q}^2(v)}{v^2 \bar{q}'(v)} = \frac{(2 - 2v^2 + v^3)^2}{2 - 6v^2 + 4v^3} \geq \lim_{v \to 0^+} \frac{(2 - 2v^2 + v^3)^2}{2 - 6v^2 + 4v^3} = 2$$

for all $v \in (0, 1)$ (since g_2 is increasing in $(0, 1)$). Hence, (4.4) holds and the DMRL class is preserved.

Example 3.3 in Navarro (2018a) shows that the DMRL class is also preserved in a parallel system with dependent components and that the IMRL class can be preserved under the formation of series systems with dependent components. Example 3.4 in this paper shows that the DMRL class is not preserved under the formation of series systems (order statistics) with IID components. Lindqvist and Samaniego (2019) proved that the IMRL class cannot be preserved in systems with IID components. More preservation properties for the IFR and ILR classes in series and parallel systems can be seen Navarro and Shaked (2010) and in the references therein.

4.3   Systems with Non-ID Components

In this case we will use the representation of the distributions of systems as generalized distortions of the distributions of the components obtained in Theorem 2.9 and the preservation results for them obtained in Navarro et al. (2014). They can be stated as follows.

Theorem 4.3 Let $F_Q = Q(F_1, \ldots, F_n)$ be a generalized distorted distribution. Then:

(i) The IFR (DFR) class is preserved by Q if $\alpha_i(\mathbf{u}) = u_i \partial_i \bar{Q}(\mathbf{u})/\bar{Q}(\mathbf{u})$ are decreasing (increasing) for $i = 1, \ldots, n$ and for $\mathbf{u} = (u_1, \ldots, u_n) \in (0, 1)^n$.

(ii) *The DRFR (IRFR) class is preserved by Q if $\bar{\alpha}_i(\mathbf{u}) = u_i \partial_i Q(\mathbf{u})/Q(\mathbf{u})$ are decreasing (increasing) for $i = 1, \ldots, n$ and for $\mathbf{u} = (u_1, \ldots, u_n) \in (0, 1)^n$.*

(iii) *The NBU (NWU) class is preserved by Q if $\bar{Q}$ is submultiplicative (supermultiplicative), that is,*

$$\bar{Q}(u_1 v_1, \ldots, u_n v_n) \leq \bar{Q}(u_1, \ldots, u_n)\bar{Q}(v_1, \ldots, v_n), \ (\geq) \qquad (4.7)$$

for all $u_1, \ldots, u_n, v_1, \ldots, v_n \in [0, 1]$.

(iv) *The IFRA (DFRA) class is preserved by Q if $\bar{Q}$ satisfies*

$$\bar{Q}(u_1^c, \ldots, u_n^c) \geq (\bar{Q}(u_1, \ldots, u_n))^c, \ (>) \qquad (4.8)$$

for all $u_1, \ldots, u_n, c \in [0, 1]$.

Proof To prove that (i) holds, we recall that the hazard rate of F_Q can be written as

$$h_Q(t) = \sum_{i=1}^{n} \alpha_i(\bar{F}_1(t), \ldots, \bar{F}_n(t))h_i(t),$$

where $h_1, \ldots, h_n$ are the hazard rate functions of $F_1, \ldots, F_n$. If $h_1, \ldots, h_n$ are increasing (decreasing) and $\alpha_1, \ldots, \alpha_n$ are decreasing (increasing), then h_Q is increasing (decreasing) since all the reliability functions are decreasing.

The proof of (ii) is similar.

To prove (iii), we note that F_Q is NBU (NWU) iff

$$\bar{Q}(\bar{F}_1(x + t), \ldots, \bar{F}_n(x + t)) \leq \bar{Q}(\bar{F}_1(x), \ldots, \bar{F}_n(x))\bar{Q}(\bar{F}_1(t), \ldots, \bar{F}_n(t)), \ (\geq)$$

for all $x, t \geq 0$. If we assume that F_i are NBU for $i = 1, \ldots, n$ and $\bar{Q}$ is submultiplicative, that is, it satisfies (4.7), then

$$\bar{Q}(\bar{F}_1(x + t), \ldots, \bar{F}_n(x + t)) \leq \bar{Q}(\bar{F}_1(x)\bar{F}_1(t), \ldots, \bar{F}_n(x)\bar{F}_n(x))$$
$$\leq \bar{Q}(\bar{F}_1(x), \ldots, \bar{F}_n(x))\bar{Q}(\bar{F}_1(t), \ldots, \bar{F}_n(t))$$

and so F_Q is NBU. The proof for the NWU class is similar.

The proof of (iv) is analogous to the preceding one.

Note that we just have sufficient conditions. We must say that the conditions in (i) and (ii) are quite strong. The conditions in (i) (in (ii)) are satisfied by series (parallel) systems with independent components and so both classes are preserved. As we can see in Fig. 3.9, the IFR and DFR classes are not preserved in parallel systems with independent components. However, the conditions for the NBU and IFRA classes given in (iii) and (iv) are mild conditions. Actually, we can prove that they holds for any coherent system with independent components. These properties were proved in Esary et al. (1970) (see also Barlow and Proschan 1975, p. 85). We include them in the next proposition with a different proof.

Proposition 4.1 (Esary et al. 1970) *If $\bar{Q}$ is the distortion function of a coherent system with independent components, then $\bar{Q}$ satisfies (4.7) and (4.8).*

Proof To prove (4.7) we note that $\bar{Q}(u_1v_1, \ldots, u_nv_n)$ is the dual distortion function of a system T_1 with the same structure as the original system but replacing its components with series systems with two independent components. Let $X_1, \ldots, X_n, Y_1, \ldots, Y_n$ be the component lifetimes in that system.

Analogously, the function

$$\bar{Q}(u_1, \ldots, u_n)\bar{Q}(v_1, \ldots, v_n)$$

is the dual distortion function of the system T_2 obtained by connecting in series two independent copies T_3 and T_4 of the original system with (independent) component lifetimes $X_1, \ldots, X_n$ and $Y_1, \ldots, Y_n$, respectively.

Thus, if $P_1, \ldots, P_r$ are the minimal path sets of the original system, the system T_1 works iff at least one $P_i = \{i_1, \ldots, i_k\}$ of these minimal path sets works, that is, the components with lifetimes $X_{i_1}, \ldots, X_{i_k}, Y_{i_1}, \ldots, Y_{i_k}$ work. Then the corresponding path sets in T_3 and T_4 work, and so T_2 works as well. Note that we have proved that T_2 works whenever T_1 works, that is, $T_1 \leq T_2$. So $T_1 \leq_{ST} T_2$ holds for any $F_1, \ldots, F_n$ and hence their dual distortion functions are ordered as in (4.7).

To prove (4.8) we first note that, when the components are independent, $\bar{Q}$ is the polynomial obtained by extending the structure function in the pivotal decomposition to $\mathbb{R}$.

We are going to prove (4.8) by induction in the number of components. It is clearly true for $n = 1$ since $\bar{Q}(u_1) = u_1$. Then, we assume that it is true for all the coherent systems with less than n independent components.

As the components are independent, $\bar{Q}(u_1, \ldots, u_n)$ is a polynomial of degree 1 in u_1 and so we have

$$\bar{Q}(u_1, \ldots, u_n) = u_1\bar{Q}(1, u_2, \ldots, u_n) + (1 - u_1)\bar{Q}(0, u_2, \ldots, u_n)$$

(since these linear functions coincide in two points $u_1 = 0$ and $u_1 = 1$).

Then

$$\bar{Q}(u_1^c, \ldots, u_n^c) = u_1^c\bar{Q}(1, u_2^c, \ldots, u_n^c) + (1 - u_1^c)\bar{Q}(0, u_2^c, \ldots, u_n^c)$$

holds for all $c \in (0, 1)$, where $\bar{Q}(1, u_2, \ldots, u_n)$ and $\bar{Q}(0, u_2, \ldots, u_n)$ are dual distortion functions of coherent systems with less than n components (we delete the first component and all the irrelevant components). Hence, by the induction hypothesis, we get

$$\bar{Q}(u_1^c, \ldots, u_n^c) \geq u_1^c(\bar{Q}(1, u_2, \ldots, u_n))^c + (1 - u_1^c)(\bar{Q}(0, u_2, \ldots, u_n))^c.$$

Finally, we use that $\lambda^c y^c + (1 - \lambda^c)x^c \geq (\lambda y + (1 - \lambda)x)^c$ holds for all $0 \leq x \leq y$ and all $0 \leq \lambda \leq 1$. Then, as $\bar{Q}(1, u_2, \ldots, u_n) \geq \bar{Q}(0, u_2, \ldots, u_n)$, we get

$$\bar{Q}(u_1^c, \ldots, u_n^c) \geq (u_1\bar{Q}(1, u_2, \ldots, u_n) + (1 - u_1)\bar{Q}(0, u_2, \ldots, u_n))^c$$

$$= (\bar{Q}(u_1, \ldots, u_n))^c.$$

and so (4.8) holds.   □

We have seen in Fig. 4.3, left, that a coherent system with IID exponential components preserves the NBU property. From the preceding proposition, this is actually the case for any coherent system with independent components. This property says that a new coherent system with IND components is always ST better than the residual lifetimes of its used systems with any age $t > 0$. However, note that, as we can see in that figure, these residual lifetimes are not ST ordered for any t (in that figure the worse used system is obtained at $t = 1.444$).

4.4   Residual and Inactivity Times of Systems

In the preceding sections we have analyzed the behavior of the residual lifetimes of a system T with age t defined as $T_t = (T - t | T > t)$ for $t \geq 0$. In these residual lifetimes we assume that, at time t, we just know that the system is working. As we have seen, the reliability function of T_t is

$$\bar{F}_T(x|t) = \frac{\bar{F}_T(t + x)}{\bar{F}_T(t)} = \frac{\bar{Q}(\bar{F}_1(t + x), \ldots, \bar{F}_n(t + x))}{\bar{Q}(\bar{F}_1(t), \ldots, \bar{F}_n(t))} \qquad (4.9)$$

for t such that $\bar{F}_T(t) > 0$, where $\bar{Q}$ is the dual distortion function of the system.

In other situations, we may have different information at time t. For example, we may know that all the components are working. In this case, the system residual lifetime is

$$T_t^* = (T - t | X_1 > t, \ldots, X_n > t)$$

for t such that $\Pr(X_1 > t, \ldots, X_n > t) > 0$. Other options will be considered later. Note that $T_t =_{ST} T_t^*$ when $T = X_{1:n}$ (series system).

One can presume that T_t^* is better than T_t. Thus, if we are in a plane, we would prefer to know that all the engines are working, instead of just to know that the plane engine system is working. However, we will see that, surprisingly, this is not always the case.

To compare the reliability functions of these residual lifetimes, we need to write them as distortions of the same distributions. To this purpose we will use the residual lifetimes of the components defined as $X_{i,t} = (X_i - t | X_i > t)$ for $i = 1, \ldots, n$ with reliability functions

$$\bar{F}_i(x|t) = \frac{\bar{F}_i(x + t)}{\bar{F}_i(t)}$$

for $x \geq 0$. They are defined for $t \geq 0$ such that $\bar{F}_i(t) > 0$ for $i = 1, \ldots, n$. We shall assume these conditions in this section whenever we consider these conditional distributions.

The representation for the first case is immediate since from (4.9) we get

$$\bar{F}_T(x|t) = \frac{\bar{Q}(\bar{F}_1(t)\bar{F}_1(x|t), \ldots, \bar{F}_n(t)\bar{F}_n(x|t))}{\bar{Q}(\bar{F}_1(t), \ldots, \bar{F}_n(t))}$$
$$= \bar{Q}_t(\bar{F}_1(x|t), \ldots, \bar{F}_n(x|t)), \qquad (4.10)$$

where
$$\bar{Q}_t(u_1, \ldots, u_n) = \frac{\bar{Q}(\bar{F}_1(t)u_1, \ldots, \bar{F}_n(t)u_n)}{\bar{Q}(\bar{F}_1(t), \ldots, \bar{F}_n(t))}$$

is a distortion function which depends on $\bar{Q}$ (i.e. the system structure and the copula of the component lifetimes) and on $\bar{F}_1(t), \ldots, \bar{F}_n(t)$ for $t > 0$.

The representation for the second case is stated in the following proposition. It is extracted from Navarro (2018c).

Proposition 4.2 *The reliability function of* $T^* = (T - t|X_1 > t, \ldots, X_n > t)$ *can be written as*
$$\bar{F}_{T^*}(x|t) = \bar{Q}_t^*(\bar{F}_1(x|t), \ldots, \bar{F}_n(x|t)), \tag{4.11}$$

where $\bar{Q}_t^*$ *is a distortion function which depends on the system structure (the minimal path sets), the survival copula* $\widehat{C}$ *and on* $\bar{F}_1(t), \ldots, \bar{F}_n(t)$.

Proof Recall that if $P_1, \ldots, P_r$ are the minimal path sets of the system, then its lifetime can be written as $T = \max_{i=1,\ldots,r} \min_{j \in P_i} X_j$. Hence, the reliability function of T_t^* is

$$\bar{F}_{T^*}(x|t) = \Pr(T - t > x | X_1 > t, \ldots, X_n > t)$$
$$= \frac{\Pr(\max_{i=1,\ldots,r} \min_{j \in P_i} X_j > t + x, X_1 > t, \ldots, X_n > t)}{\Pr(X_1 > t, \ldots, X_n > t)}$$
$$= \frac{N(t)}{D(t)},$$

where
$$D(t) = \Pr(X_1 > t, \ldots, X_n > t) = \widehat{C}(\bar{F}_1(t), \ldots, \bar{F}_n(t)) > 0.$$

This implies $\bar{F}_i(t) > 0$ for $i = 1, \ldots, n$. Then, by applying the inclusion-exclusion formula to the numerator $N(t)$ above, we obtain

$$N(t) = \sum_{i=1}^{r} \Pr\left(\min_{j \in P_i} X_j > t + x, X_1 > t, \ldots, X_n > t\right)$$
$$- \sum_{i=1}^{r-1} \sum_{j=i+1}^{r} \Pr\left(\min_{k \in P_i \cup P_j} X_k > t + x, X_1 > t, \ldots, X_n > t\right)$$
$$+ \cdots + (-1)^{r+1} \Pr\left(\min_{k \in P_1 \cup \cdots \cup P_r} X_k > t + x, X_1 > t, \ldots, X_n > t\right),$$

where

$$\bar{G}_P(t) := \Pr\left(\min_{j \in P} X_j > t + x, X_1 > t, \ldots, X_n > t\right)$$
$$= \Pr\left(\cap_{j \in P} X_j > t + x, \cap_{j \notin P} X_j > t\right)$$
$$= \widehat{C}_P(\bar{F}_1(x|t), \ldots, \bar{F}_n(x|t)) \tag{4.12}$$

and
$$\widehat{C}_P(u_1, \ldots, u_n) = \widehat{C}(u_1^P, \ldots, u_n^P)$$
with $u_i^P := u_i \bar{F}_i(t)$ if $i \in P$ and $u_i^P := \bar{F}_i(t)$ if $i \notin P$. Therefore

$$N(t) = \sum_{i=1}^{r} \bar{G}_{P_i}(t) - \sum_{i=1}^{r-1} \sum_{j=i+1}^{r} \bar{G}_{P_i \cup P_j}(t) + \cdots + (-1)^{r+1} \bar{G}_{P_1 \cup \cdots \cup P_r}(t).$$

Hence, by using (4.12), (4.11) holds. $\square$

If the components are ID, then the above representations for T_t and T_t^* can be reduced to univariate distortions. In both cases, we can use the results for distorted distributions stated in the preceding chapter to compare these residual lifetimes. Let us see a simple example.

Example 4.6 We consider a parallel system $T = \max(X_1, X_2)$ with two possible dependent component with lifetimes X_1 and X_2 and with survival copula $\widehat{C}$. As we have seen before, the system reliability function can be written as

$$\bar{F}_T(t) = \bar{Q}(\bar{F}_1(t), \bar{F}_2(t)),$$

where $\bar{Q}(u_1, u_2) = u_1 + u_2 - \widehat{C}(u_1, u_2)$ for $u_1, u_2 \in [0, 1]$.

Let T_t and T_t^* be the residual lifetimes defined above. Then the reliability function of T_t can be written as in (4.10) for

$$\bar{Q}_t(u_1, u_2) = \frac{u_1 \bar{F}_1(t) + u_2 \bar{F}_2(t) - \widehat{C}(u_1 \bar{F}_1(t), u_2 \bar{F}_2(t))}{\bar{F}_1(t) + \bar{F}_2(t) - \widehat{C}(\bar{F}_1(t), \bar{F}_2(t))}.$$

Analogously, the reliability function of T_t^* can be written as in (4.11) for

$$\bar{Q}_t^*(u_1, u_2) = \frac{\widehat{C}(u_1 \bar{F}_1(t), \bar{F}_2(t)) + \widehat{C}(\bar{F}_1(t), u_2 \bar{F}_2(t)) - \widehat{C}(u_1 \bar{F}_1(t), u_2 \bar{F}_2(t))}{\widehat{C}(\bar{F}_1(t), \bar{F}_2(t))}.$$

If the components are independent, that is, $\widehat{C}(u_1, u_2) = u_1 u_2$, then

$$\bar{Q}_t(u_1, u_2) = \frac{u_1 \bar{F}_1(t) + u_2 \bar{F}_2(t) - u_1 u_2 \bar{F}_1(t) \bar{F}_2(t)}{\bar{F}_1(t) + \bar{F}_2(t) - \bar{F}_1(t) \bar{F}_2(t)}$$

and

$$\bar{Q}_t^*(u_1, u_2) = \frac{u_1 \bar{F}_1(t) \bar{F}_2(t) + u_2 \bar{F}_1(t) \bar{F}_2(t) - u_1 u_2 \bar{F}_1(t) \bar{F}_2(t)}{\bar{F}_1(t) \bar{F}_2(t)} = \bar{Q}(u_1, u_2).$$

In particular, if we assume that the components are ID with a common reliability $\bar{F}$ and a survival copula $\widehat{C}$, then these representations can be reduced to $\bar{F}_T(x|t) = \bar{q}_t(\bar{F}(x|t))$ and $\bar{F}_{T^*}(x|t) = \bar{q}_t^*(\bar{F}(x|t))$, where $\bar{F}(x|t) = \bar{F}(x+t)/\bar{F}(t)$,

$$\bar{q}_t(u) = \frac{2u\bar{F}(t) - \widehat{C}(u\bar{F}(t), u\bar{F}(t))}{2\bar{F}(t) - \widehat{C}(\bar{F}(t), \bar{F}(t))}$$

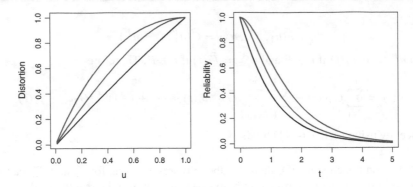

Fig. 4.5 Dual distortion functions (left) and reliability functions (right) for T (red), T_t (black) and T_t^* (blue) for the parallel system in Example 4.6 when $t = 1$, the components have a common standard exponential distribution and a Clayton–Oakes survival copula

and

$$\bar{q}_t^*(u) = \frac{\widehat{C}(u\bar{F}(t), \bar{F}(t)) + \widehat{C}(\bar{F}(t), u\bar{F}(t)) - \widehat{C}(u\bar{F}(t), u\bar{F}(t))}{\widehat{C}(\bar{F}(t), \bar{F}(t))}$$

for $u \in [0, 1]$. By comparing (or plotting) these functions we can see if $T_t \leq_{ST} T_t^*$ holds (as expected). For example, in the IID case we get

$$\bar{q}_t(u) = u\frac{2 - u\bar{F}(t)}{2 - \bar{F}(t)} \leq 2u - u^2 = \bar{q}_t^*(u)$$

for all $u \in [0, 1]$, and so $T_t \leq T_t^*$ for any t and any $\bar{F}$.

If the components are dependent with the Clayton–Oakes survival copula in (3.12), and $c = \bar{F}(t) = \exp(-1)$, we obtain the plots in Fig. 4.5, left. As they are ordered, we have $T_t \leq_{ST} T_t^*$. In Fig. 4.5, right we can see the respective reliability functions when the components have a common standard exponential distribution. We can also compare them with the original system T (red curves). Note that T is NBU and so $T_t \leq_{ST} T$ holds for all t (and, in particular, for $t = 1$). However, we have $T \leq_{ST} T_t^*$ for $t = 1$. It can be proved analytically (see Navarro 2018c) that $T_t \leq_{ST} T_t^*$ is a general property for any t (any c) and any Clayton–Oakes copula with positive correlation. Therefore, for these copulas, $T_t \leq_{ST} T_t^*$ holds for any t and any $\bar{F}$.

However, if we choose the following Gumbel-Barnett Archimedean copula

$$\widehat{C}(u_1, u_2) = u_1 u_2 \exp[-(\ln u_1)(\ln u_2)],$$

then we obtain the plots in Fig. 4.6. There we can see that T_t and T_t^* are not ST ordered for $t = 1$. However, note that both are worse than T (red lines). The code to get these plots is the following. For other t values, copulas or distributions just change the corresponding definitions in that code.

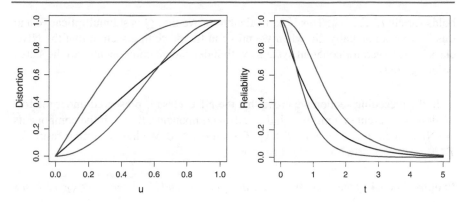

Fig. 4.6 Dual distortion functions (left) and reliability functions (right) for T (red), T_t (black) and T_t^* (blue) for the parallel system in Example 4.6 when $t = 1$, the components have a common standard exponential distribution and a Gumbel-Barnett survival copula

```
# Figure 4.6
R<-function(x) exp(-x)
t<-1
c<-R(t)
C<-function(u,v) u*v*exp(-log(u)*log(v))
q<-function(u) 2*u-C(u,u)
qt<-function(u) (2*u*c-C(u*c,u*c))/(2*c-C(c,c))
qt2<-function(u) (2*C(u*c,c)-C(u*c,u*c))/C(c,c)
curve(qt(x),ylab='Distortion',xlab='u',lwd=2)
curve(q(x),add=T,col='red',lwd=2)
curve(qt2(x),add=T,col='blue',lwd=2)
curve(qt(R(x)),ylab='Reliability',0,5,xlab='t',lwd=2)
curve(q(R(x)),add=T,col='red',lwd=2)
curve(qt2(R(x)),add=T,col='blue',lwd=2)
```

We can use the equality $\bar{Q}_t^* = \bar{Q}$ to prove that $T_t \leq_{ST} T_t^*$ holds for any t when the components are independent. This result was obtain in Pellerey and Petakos (2002) (with a different proof).

Proposition 4.3 *If the component lifetimes are independent, then* $\bar{Q}_t^* = \bar{Q}$ *and* $T_t \leq_{ST} T_t^*$ *hold for any* t

Proof When the components are independent, $\bar{Q}$ is a multinomial and so $\bar{Q}_t^* = \bar{Q}$ (see the proof of Proposition 4.2). Hence, from the representations (4.10) and (4.11), $T_t \leq_{ST} T_t^*$ holds iff

$$\bar{Q}_t(u_1, \ldots, u_n) = \frac{\bar{Q}(\bar{F}_1(t)u_1, \ldots, \bar{F}_n(t)u_n)}{\bar{Q}(\bar{F}_1(t), \ldots, \bar{F}_n(t))} \leq \bar{Q}(u_1, \ldots, u_n)$$

holds for all $u_1, \ldots, u_n$. This is equivalent to prove that $\bar{Q}$ is submultiplicative and this is what we actually did when we prove in the preceding section that the NBU class is preserved for coherent systems with independent components. So the stated ordering holds.   □

In the preceding section we prove that the NBU class is preserved under the formation of coherent systems with independent components, that is, if the components are NBU, then so is T, that is, $T \geq_{ST} T_t$ for any $t \geq 0$. We have a similar result for T_t^*.

Proposition 4.4 *If the components are independent and NBU, then $T \geq_{ST} T_t^*$ holds for any t.*

Proof If the components are NBU, then $\bar{F}_i(x) \geq \bar{F}_i(x|t)$ for any $t \geq 0$ and $i = 1, \ldots, n$. Hence, as $\bar{Q}_t^* = \bar{Q}$, we have

$$\begin{aligned}
\bar{F}_{T*}(x|t) &= \bar{Q}_t^*(\bar{F}_1(x|t), \ldots, \bar{F}_n(x|t)) \\
&= \bar{Q}(\bar{F}_1(x|t), \ldots, \bar{F}_n(x|t)) \\
&\leq \bar{Q}(\bar{F}_1(x), \ldots, \bar{F}_n(x)) \\
&= \bar{F}_T(x).
\end{aligned}$$ □

As an immediate consequence we have that, if the components have exponential distributions, then T_t^* has the lack of memory property, that is, $T =_{ST} T_t^* \geq_{ST} T_t$ for all t. If they are just independent and NBU (or IFR), then $T \geq_{ST} T_t^* \geq_{ST} T_t$ for all t.

The last example shows that these properties cannot be extended to the HR order.

Example 4.7 Let us consider a parallel system with two independent components. Its distortion function is

$$\bar{Q}(u_1, u_2) = u_1 + u_2 - u_1 u_2$$

for $u_1, u_2 \in [0, 1]$. Analogously, the distortion functions of T_t and T_t^* are

$$\bar{Q}_t(u_1, u_2) = \frac{u_1 \bar{F}_1(t) + u_2 \bar{F}_2(t) - u_1 u_2 \bar{F}_1(t) \bar{F}_2(t)}{\bar{F}_1(t) + \bar{F}_2(t) - \bar{F}_1(t) \bar{F}_2(t)}$$

and $\bar{Q}_t^* = \bar{Q}$. A straightforward calculation shows that $\bar{Q} \geq \bar{Q}_t$ for all t. Hence, as stated in Proposition 4.3, $T_t \leq_{ST} T_t^*$ holds for all t and all $\bar{F}_1, \bar{F}_2$. For example, in Fig. 4.7, left, we can see that the respective reliability functions are ordered. We do not include the reliability function of T since it coincides with that of T_t^* (blue line) for all $t \geq 0$. In the right plot we can see that the respective hazard rate functions are not ordered, that is, $T_t \leq_{HR} T_t^*$ does not hold. This is a very surprising property! Actually note that, after some time, T_t^* will be worse than T_t. The code to get these plots is the following:

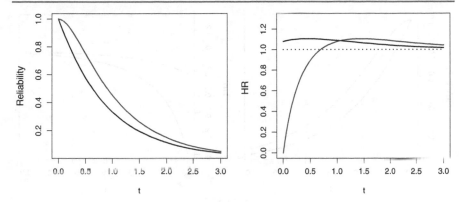

Fig. 4.7 Reliability functions (left) and hazard rate functions (right) for T_t (black) and T_t^* (blue) for the parallel system in Example 4.7 when $t = 1$ and the components are independent having exponential distributions with hazard rates 1 (dashed line, right) and 2

```
# Reliability functions
R1<-function(x) exp(-x)
R2<-function(x) exp(-2*x)
t<-1
c1<-R1(t)
c2<-R2(t)
Q<-function(u,v) u+v-u*v
Qt<-function(u,v) (u*c1+v*c2-u*v*c1*c2)/(c1+c2-c1*c2)
curve(Qt(R1(x),R2(x)),xlab='t',ylab='Reliability',0,3,lwd=2)
curve(Q(R1(x),R2(x)),add=T,col='blue',lwd=2)

# Hazard rate functions
f1<-function(x) exp(-x)
f2<-function(x) 2*exp(-2*x)
Q1<-function(u,v) 1-v
Q2<-function(u,v) 1-u
Qt1<-function(u,v) (c1-v*c1*c2)/(c1+c2-c1*c2)
Qt2<-function(u,v) (c2-u*c1*c2)/(c1+c2-c1*c2)
f<-function(t) f1(t)*Q1(R1(t),R2(t))+f2(t)*Q2(R1(t),R2(t))
ft<-function(t) f1(t)*Qt1(R1(t),R2(t))+f2(t)*Qt2(R1(t),R2(t))
curve(ft(x)/Qt(R1(x),R2(x)),xlab='t',ylab='HR',0,3,
ylim=c(0,1.3),lwd=2)
curve(f(x)/Q(R1(x),R2(x)),add=T,col='blue',lwd=2)
curve(x+1-x,lty=3,add=T,lwd=2)
```

However, if we assume that the components are IID, then the respective distortion functions are

$$\bar{q}(u) = 2u - u^2,$$

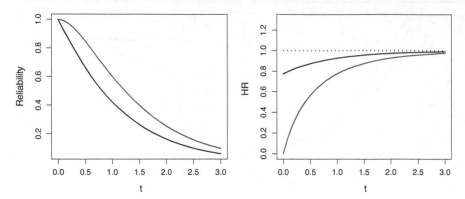

Fig. 4.8 Reliability functions (left) and hazard rate functions (right) for T_t (black) and T_t^* (blue) for the parallel system in Example 4.7 when $t = 1$ and the components are IID having a common exponential distribution with hazard rate 1 (dashed line, right)

$$\bar{q}_t(u) = \frac{2u - u^2 \bar{F}(t)}{2 - \bar{F}(t)}$$

and $\bar{q}_t^* = \bar{q}$ for all $t \geq 0$. As

$$\frac{\bar{q}_t^*(u)}{\bar{q}_t(u)} = (2 - \bar{F}(t)) \frac{2 - u}{2 - u \bar{F}(t)}$$

is decreasing in u (for all t), then we have $T_t \leq_{HR} T_t^*$ for all t and all $\bar{F}$. In Fig. 4.8, we can see the respective reliability functions (left) and hazard rate functions (right) when $t = 1$ for a standard exponential distribution (they can be plotted with the code written above). As we can see, in both cases, the limit behaviors of the hazard rate functions coincide with that of the hazard rate of the strongest component.

We can go further and as

$$\frac{(\bar{q}_t^*)'(u)}{\bar{q}_t'(u)} = (2 - \bar{F}(t)) \frac{1 - u}{1 - u \bar{F}(t)}$$

is decreasing in u (for all t), then we have $T_t \leq_{LR} T_t^*$ for all t and all absolutely continuous $\bar{F}$.

Other residual lifetimes and inactivity times can be explored in a similar way. For example, we can consider the residual lifetime

$$(T - t | X^I \leq t, X_J > t),$$

where $I, J \subseteq [n]$, $I \cap J = \emptyset$, $X^I = \max_{i \in I} X_i$, $X_J = \min_{i \in J} X_i$ and we assume that the event $\{X^I \leq t, X_J > t\}$ implies $\{T > t\}$. Conversely, if it implies $\{T \leq t\}$ then we can consider the inactivity time

$$(t - T | X^I \leq t, X_J > t).$$

We can also consider the residual lifetime

$$(T - t | X_i = t_1, X^I \leq t, X_J > t)$$

for $t_1 \leq t$ when the event implies $T > t$ and other similar options in an interval $[t_1, t_2]$. These options were explored in Navarro and Durante (2017), Navarro et al. (2017), Navarro (2018c) and Navarro and Calì (2019). Signature representations for used systems' residual lifetimes can be seen in Navarro et al. (2008).

4.5   Limiting Behavior

As we have seem in preceding examples, in many cases, the limiting behavior when the time goes on of the hazard rate function of the system is similar to that of the components in the ID case. This is due to similar properties of mixtures and generalized mixtures and to the representation based on minimal path sets (which proves that the system distribution is a generalized mixture of series systems). However, this is not always the case. In this section we explore these properties in order to determine the limiting behavior of reliability, hazard rate and mean residual life functions of the system. These properties are based on the following results for distorted distributions extracted from Burkschat and Navarro (2018).

Proposition 4.5 *Let $F_1 = q_1(F)$ and $F_2 = q_2(F)$ be two distorted distributions from the same distribution function F with $\bar{F}(t) > 0$ for all t. Let $\bar{q}_1$ and $\bar{q}_2$ be the associated dual distortion functions.*

(i) *The reliability functions satisfy*

$$\lim_{t \to \infty} \frac{\bar{F}_1(t)}{\bar{F}_2(t)} = \lim_{u \to 0^+} \frac{\bar{q}_1(u)}{\bar{q}_2(u)}.$$

(ii) *The hazard rate functions satisfy*

$$\lim_{t \to \infty} \frac{h_1(t)}{h_2(t)} = \lim_{u \to 0^+} \frac{\alpha_1(u)}{\alpha_2(u)},$$

where $\alpha_i(u) = u\bar{q}_i'(u)/\bar{q}_i(u)$ for $u \in [0, 1]$ and $i = 1, 2$.

(iii) *The mean residual life functions satisfy*

$$\lim_{t \to \infty} \frac{m_1(t)}{m_2(t)} = 1$$

whenever $\lim_{u \to 0^+} \bar{q}_1(u)/\bar{q}_2(u) = c > 0$.

(iv) *The pdf satisfy*

$$\lim_{t \to \infty} \frac{f_1(t)}{f_2(t)} = \lim_{u \to 0^+} \frac{\bar{q}_1'(u)}{\bar{q}_2'(u)}.$$

(v) *The hazard rate functions satisfy*

$$\lim_{t \to \infty} \frac{h_1(t)}{h_2(t)} = 1$$

whenever $\lim_{u \to 0^+} \bar{q}_1'(u)/\bar{q}_2'(u) = c > 0$.

Proof The proofs of (i), (ii) and (iv) are immediate from (2.32), (2.33) and (2.34). To prove (iii), we note that

$$\lim_{t\to\infty}\frac{m_1(t)}{m_2(t)}=\lim_{t\to\infty}\frac{\bar{F}_2(t)}{\bar{F}_1(t)}\frac{\int_t^\infty \bar{F}_1(x)dx}{\int_t^\infty \bar{F}_2(x)dx}=\frac{1}{c}\lim_{t\to\infty}\frac{\int_t^\infty \bar{F}_1(x)dx}{\int_t^\infty \bar{F}_2(x)dx}$$

since $c > 0$ and (i) holds. Moreover, from L'Hôpital's rule, we have

$$\lim_{t\to\infty}\frac{\int_t^\infty \bar{F}_1(x)dx}{\int_t^\infty \bar{F}_2(x)dx}=\lim_{t\to\infty}\frac{\bar{F}_1(t)}{\bar{F}_2(t)}=c.$$

Therefore (iii) holds. The proof of (v) is similar. $\square$

Clearly, the preceding proposition can be applied to compare the behavior of systems with ID components. In particular, it can also be applied to compare a system with its ID components (by choosing $\bar{q}_2(u) = u$ and $\alpha_2(u) = 1$ for $u \in [0, 1]$). This fact can be used to detect situations where the system behavior coincides with that of its ID components. In other cases, it will coincide with the behavior of the functions of some series systems. Let us see an example.

Example 4.8 Let us consider the coherent systems with lifetimes

$$T_1 = \min(X_1, \max(X_2, X_3))$$

and

$$T_2 = \max(X_1, \min(X_2, X_3)).$$

Let us assume that the components are IID$\sim F$. Then the respective distortion functions are

$$\bar{q}_1(u) = 2u^2 - u^3$$

and

$$\bar{q}_2(u) = u + u^2 - u^3$$

for $u \in [0, 1]$. Clearly, the behavior of the reliability function of the second system is similar to that of the components since

$$\lim_{t\to\infty}\frac{\bar{F}_{T_2}(t)}{\bar{F}(t)}=\lim_{u\to 0^+}\frac{\bar{q}_2(u)}{u}=\lim_{u\to 0^+}\frac{u+u^2-u^3}{u}=1$$

for any F. As a consequence, from (iii) in the preceding proposition, we have that the MRL functions also have a similar behavior, that is,

$$\lim_{t\to\infty}\frac{\bar{m}_{T_2}(t)}{m(t)}=1.$$

For the HR functions we obtain

$$\lim_{t\to\infty}\frac{h_{T_2}(t)}{h(t)}=\lim_{u\to 0^+}\alpha_2(u)=\lim_{u\to 0^+}\frac{u\bar{q}_2'(u)}{\bar{q}_2(u)}=\lim_{u\to 0^+}u\frac{1+2u-3u^2}{u+u^2-u^3}=1.$$

This property can also be obtained from Proposition 4.5, (v).

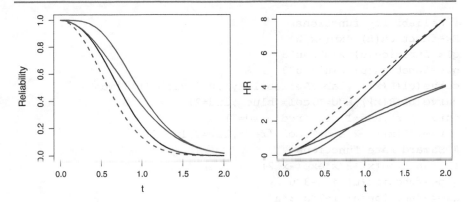

Fig. 4.9 Reliability functions (left) and hazard rate functions (right) for T_1 (black) and T_2 (blue) for the systems in Example 4.8 when the components are IID with a Weibull distribution with hazard rate $2t$ (continuous red line). The dashed red lines represent the functions of the series system $X_{1:2}$

Analogously, for the first system we have

$$\lim_{t\to\infty} \frac{h_{T_1}(t)}{h(t)} = \lim_{u\to 0^+} \alpha_1(u) = \lim_{u\to 0^+} \frac{u\bar{q}_1'(u)}{\bar{q}_1(u)} = \lim_{u\to 0^+} u\frac{4u - 3u^2}{2u^2 - u^3} = 2.$$

However, for the reliability functions we have

$$\lim_{t\to\infty} \frac{\bar{F}_{T_1}(t)}{\bar{F}(t)} = \lim_{u\to 0^+} \frac{\bar{q}_1(u)}{u} = \lim_{u\to 0^+} \frac{2u^2 - u^3}{u} = 0$$

(the system reliability goes faster to zero that that of the components). To get an equivalent system we must consider the series system with two components $X_{1:2} = \min(X_1, X_2)$ with reliability $\bar{F}_{1:2}(t) = \bar{F}^2(t)$ and then

$$\lim_{t\to\infty} \frac{\bar{F}_{T_1}(t)}{\bar{F}_{1:2}(t)} = \lim_{u\to 0^+} \frac{\bar{q}_1(u)}{u^2} = \lim_{u\to 0^+} \frac{2u^2 - u^3}{u^2} = 2.$$

Hence, from Proposition 4.5, (iii) and (v), we also have

$$\lim_{t\to\infty} \frac{h_{T_1}(t)}{h_{1:2}(t)} = \lim_{t\to\infty} \frac{m_{T_1}(t)}{m_{1:2}(t)} = 1$$

for any F. Similar results can be obtained for dependent components with a given copula.

To illustrate these results we plot the reliability and hazard rate functions of these systems in Fig. 4.9 when the components are ID with a common Weibull reliability $\bar{F}(t) = \exp(-t^2)$ for $t \geq 0$. The hazard rate of the components is $h(t) = 2t$ (continuous red line) and that of $X_{1:2}$ is $h_{1:2}(t) = 4t$ (dashed red line) for $t \geq 0$. The code is the following:

```
# Reliability functions
R<-function(x) exp(-x^2)
q1<-function(u) 2*u^2-u^3
q2<-function(u) u+u^2-u^3
curve(q1(R(x)),ylab='Reliability',0,2,xlab='t',lwd=2)
curve(q2(R(x)),add=T,col='blue',lwd=2)
curve(R(x),add=T,col='red',lwd=2)
curve((R(x))^ 2,add=T,col='red',lty=2,lwd=2)
# Hazard rate functions
f<-function(x) 2*x*exp(-x^2)
q1p<-function(u) 4*u-3*u^2
q2p<-function(u) 1+2*u-3*u^2
curve(f(x)*q1p(R(x))/q1(R(x)),ylab='HR',0,2,xlab='t',lwd=2)
curve(f(x)*q2p(R(x))/q2(R(x)),add=T,col='blue',lwd=2)
curve(f(x)/R(x),add=T,col='red',lwd=2)
curve(2*f(x)/R(x),add=T,col='red',lty=2,lwd=2)
```

Remark 4.2 It is easy to prove that if a system with IID components has the minimal signature $\mathbf{a} = (0, \ldots, 0, a_i, \ldots, a_n)$ with $a_i \neq 0$, then the behavior of the system aging functions is similar to that of $X_{1:i}$. In particular, we have $\lim_{t\to\infty} \bar{F}_T(t)/\bar{F}(t) = a_i$, $\lim_{t\to\infty} h_T(t)/h(t) = i$ and $\lim_{t\to\infty} m_T(t)/m_{1:i}(t) = 1$.

4.6 Bounds

A similar technique can be used to obtain bounds for these functions. The results can be stated as follows.

Proposition 4.6 *Let $F_1 = q_1(F)$ and $F_2 = q_2(F)$ be two distorted distributions from the same distribution function F. Let $\bar{q}_1$ and $\bar{q}_2$ be the associated dual distortion functions.*

(i) The reliability functions satisfy

$$\inf_{u\in(0,1]} \frac{\bar{q}_1(u)}{\bar{q}_2(u)} \leq \frac{\bar{F}_1(t)}{\bar{F}_2(t)} \leq \sup_{u\in(0,1]} \frac{\bar{q}_1(u)}{\bar{q}_2(u)}$$

for all t and all F.

(ii) The hazard rate functions satisfy

$$\inf_{u\in(0,1]} \frac{\alpha_1(u)}{\alpha_2(u)} \leq \frac{h_1(t)}{h_2(t)} \leq \sup_{u\in(0,1]} \frac{\alpha_1(u)}{\alpha_2(u)}$$

for all t and all F.

The proofs are immediate from (2.32) and (2.34).

For example, for the systems considered in the preceding example we have

$$0 \leq \bar{F}_1(t) \leq \bar{F}(t),$$

$$\bar{F}(t) \leq \bar{F}_2(t) \leq 1.25\bar{F}(t),$$

$$h(t) \leq h_1(t) \leq 2h(t) = h_{1:2}(t)$$

and

$$0 \leq h_2(t) \leq 1.105573\,h(t)$$

for any t and any F.

The bounds for the case of non-ID components can be obtained from the minimal path set representation and from the results given in Miziuła and Navarro (2018). They are based on the average reliability function

$$\bar{G} = \frac{\bar{F}_1 + \cdots + \bar{F}_n}{n}.$$

For example, if the components are independent, then we need to get bounds for the function

$$D(u_1, \ldots, u_n) = n\frac{\bar{Q}(u_1, \ldots, u_n)}{(u_1 + \cdots + u_n)}$$

and, as $\bar{Q}$ is linear in u_i for all i, these bounds are always attained at 0 or 1 for each u_i.

Problems

1. Plot the hazard rate function of a mixture of exponential distributions.
2. Plot the MRL function of a mixture of exponential distributions.
3. Prove that the mixture of two DFR distributions is also DFR.
4. Prove that the mixture of two IFR distributions is not necessarily IFR.
5. Study if the IFR/DFR classes are preserved in a system with IID components. Plot the hazard rate functions of the system for different distributions.
6. Study if the DRFR class is preserved in a system with IID components. Plot the reversed hazard rate functions of the system for different distributions.
7. Study if the NBU/NWU classes are preserved in a system with IID components. Plot the reliability functions of the residual lifetimes of the system for different ages.
8. Study if the IFR/DFR classes are preserved in a system with ID dependent components. Plot the hazard rate functions of the system for different distributions.
9. Study if the DMRL/IMRL classes are preserved in a system with IID components.
10. Study if the IFR/DFR classes are preserved in a system with IND components. Plot the hazard rate functions of the system for different distributions.
11. Study if the DRFR class is preserved in a system with IND components. Plot the reversed hazard rate functions of the system for different distributions.

12. Confirm that NBU class is preserved in a system with IND components. Plot the reliability functions of the residual lifetimes of the system for different ages.
13. Study if the IFR/DFR classes are preserved in a system with DNID components. Plot the hazard rate functions of the system for different distributions.
14. Find a system with dependent components where the NBU class is not preserved.
15. Compare the residual lifetimes T_t and T_t^* of two systems with IID components.
16. Compare the residual lifetimes T_t and T_t^* of two systems with DID components.
17. Study the limiting behavior of the aging functions of a system with IID components.
18. Study the limiting behavior of the aging functions of a system with DID components.
19. Obtain bounds for the reliability and hazard rate functions of a system with IID components.
20. Obtain bounds for the reliability and hazard rate functions of a system with DID components.

Redundancy and Repair Properties

5

Abstract

The term "redundancy" refers to the way a system can work even when some components have failed. All the coherent systems except the series systems have redundancy mechanisms in their structure functions. Moreover, sometimes, we may try to improve the reliability of a given system by adding some redundant components at different critical points. Other popular redundancy options are to add standby components in the system to replace the failed components or to repair these failed components. The main questions analyzed in this chapter are: What is the reliability of the (new) redundant system? What are the best points in the structure to add the redundant components? Which one is the best redundancy option? We also study some component importance indices that can be used to determine the best replacement options.

5.1 Redundancy Options

There are several redundancy options. Not all of them are available in practice for all the systems. Thus, we cannot use the same options for a plane or a rocket, that the ones used for ships or cars. For example, in the first cases we cannot wait for the system failure to apply the redundancy options (repairs).

In this introductory book we just analyze the most popular ones. There are two main options called "hot" and "cold" redundancies.

In the first case (**hot redundancy**), one "spare" is added to a component in the system with a given structure (which improves the behavior at this point). Both units work at the same time. The same can be done in other components as well. The most popular option is to add a new (similar) independent unit in parallel to a given component. In this case, the life length of the resulting structure at the ith position is $Y_i = \max(X_i, X_i')$, where X_i is the lifetime of the original unit and X_i' the one of the associated spare. For example, if we consider a series system with two components,

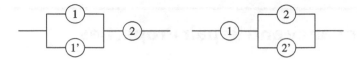

Fig. 5.1 A series system with two components and hot redundancies at positions 1 (left) and 2 (right)

it can be improved by adding a redundant component in parallel at positions 1 or 2 (see Fig. 5.1). Which one is the best option? To answer this question we need to know the characteristics of the units and the spares. Thus, we may assume that the spares have the same distributions as the units, that is, if $F_1, \ldots, F_n$ are the distribution functions of the components, we can assume that the spare at the ith position has distribution F_i. This is a reasonable assumption when the components are different. Another option is to assume that the components are similar and that a spare with distribution G can be added at any point. If the components are identically distributed and we assume $G = F_1 = \cdots = F_n$ both options coincide.

In the second case (**cold redundancy**), the spares are in standby and they replace the components when they fail. Here we also have several options in practice. For example, the standby units might be placed at fixed positions. Thus, if a plain has four engines (two in each wing), it could fly just with two (one in each wing), working the others just in case of the respective failures. Note that in a hot redundancy, the four engines are working from the beginning while in a cold redundancy the two engines in each wing work consecutively (one after the other). Which one is the best option? In other options, we might have just a spare that can be placed at any position in the system. Thus the spare wheel in a car (or a truck), can be placed at any position in case of failure.

In both options, we can consider different assumptions for the spares as well. As above, we can assume that the spares have the same distributions as the original components when they are new (because they are not working). This option is called **perfect repair** since it is equivalent to complete a perfect repair of that unit (a quite unrealistic situation in some systems). Another popular option is to assume that the spares have the same distributions of the original units but that they have the same age as the failed units. This situation is also unrealistic but it is stochastically equivalent to repair the unit to be as it was just before its failure. So it is called **minimal repair** and, in this way, in some situations, it is more realistic than the perfect repair considered above (which it is not a repair but a replacement with a new unit). In both cases, the lifetime of the mechanism at the ith position is $Y_i = X_i + X_i'$. In a perfect repair, we can assume that X_i and X_i' are independent and then the distribution of Y_i is the convolution of F_i and F_i' (see below). However, in a minimal repair, they are dependent since the distribution of X_i' depends on the age $t = X_i$ of the failed component (see below).

Finally, we note that the redundancies can be applied at different levels. If they are applied as considered above, we say that they are redundancies at the **components' level**. However, if they are applied to the entire system, then we say that they are

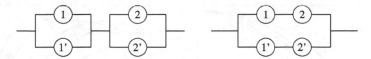

Fig. 5.2 A series system with hot redundancies at the components' level (left) and at the system's level (right)

redundancies at the **system's level**. Even more, if the system is composed of different modules with several units inside each module, the redundancy could be also applied at the **modules' level** (see, e.g., Torrado et al. 2021). For example, if we consider again a series system with two components, then we could add two spares at the components' level obtaining the system lifetime

$$T_c = \min(\max(X_1, X_1'), \max(X_2, X_2')).$$

or at the system's level obtaining

$$T_s = \max(\min(X_1, X_1'), \min(X_2, X_2')).$$

The different options can be seen in Fig. 5.2. Which one is the best option?

Many of these replacement options can be represented in a unified way by using distortions. The definition (extracted from Navarro and Fernández-Martínez 2021) is the following.

Definition 5.1 We say that $\bar{q} : [0, 1] \rightarrow [0, 1]$ is a **redundancy dual distortion function** if $\bar{q}$ is continuous, increasing and satisfies $\bar{q}(0) = 0$, $\bar{q}(1) = 1$, and $\bar{q}(u) \geq u$ for all $u \in [0, 1]$.

The purpose is to represent the reliability of the resulting mechanism at the ith position with $\bar{q}(\bar{F}_i)$ where $\bar{F}_i$ is the reliability of the original ith component. Thus, the meaning of the new condition $\bar{q}(u) \geq u$ for all $u \in [0, 1]$ is that the redundancy mechanism improves (in the stochastic order) the original one. Other additional conditions will be considered later.

Let us see some mechanisms that can be represented in this way. The first one is a hot spare connected in parallel. As mentioned above, the resulting structure at the ith position is $Y_i = \max(X_i, X_i')$ and its reliability $\bar{F}_{Y_i}(t) = \Pr(Y_i > t)$ is

$$\bar{F}_{Y_i}(t) = \Pr(\max(X_i, X_i') > t) = \Pr(X_i > t) + \Pr(X_i' > t) - \Pr(X_i > t, X_i') > t)$$

for all t. If we assume that X_i and X_i' are IID with a common reliability $\bar{F}_i$, then

$$\bar{F}_{Y_i}(t) = 2\bar{F}_i(t) - \bar{F}_i^2(t) = \bar{q}_{2:2}(\bar{F}_i(t)),$$

where $\bar{q}_{2:2}(u) = 2u - u^2$ is a distortion function satisfying $\bar{q}_{2:2}(u) \geq u$ for all $u \in [0, 1]$ (since $X_{2:2} \geq X_1$).

We can consider several changes in this model. For example, we could consider that the spare has a different (usually worse) reliability with a proportional hazard rate, i.e., $\Pr(X_i' > t) = \bar{F}_i^\theta(t)$ for $\theta > 0$, then

$$\bar{F}_{Y_i}(t) = \bar{F}_i(t) + \bar{F}_i^\theta(t) - \bar{F}_i^{1+\theta}(t) = \bar{q}_\theta(\bar{F}_i(t)),$$

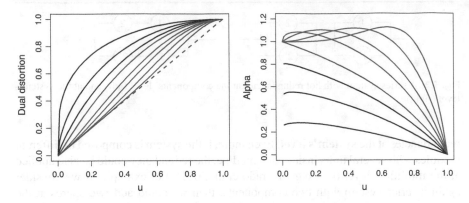

Fig. 5.3 Plots of $\bar{q}_\theta$ (left) and α_θ (right) for an independent hot redundant component with proportional hazard rate with $\theta = 0.25, 0.5, 0.75$ (black), $\theta = 1$ (red, IID case) and $\theta = 1.5, 2, 3, 5$ (blue)

where $\bar{q}_\theta(u) = u + u^\theta - u^{1+\theta}$ is a distortion function satisfying $\bar{q}_\theta(u) \geq u$ for all $u \in [0, 1]$ (since $\max(X_i, X_i') \geq X_i$). The hot redundant component is worse (better) than the original component when $\theta > 1$ ($0 < \theta < 1$). The different distortion functions can be seen in Fig. 5.3, left. Note that the IID case is obtained when $\theta = 1$ (red curve) and that they are ST ordered. As it is a distortion, its hazard rate can be written as

$$h_\theta(t) = \alpha_\theta(\bar{F}(t))h(t),$$

where $h = f/\bar{F}$ is the hazard rate of $\bar{F}$ and

$$\alpha_\theta(u) = \frac{1 + \theta u^{\theta-1} - (1+\theta)u^\theta}{1 + u^{\theta-1} - u^\theta}$$

for $u \in [0, 1]$. The plots of α_θ can be seen in Fig. 5.3, right. As they are ordered for $0 < \theta < 1$, the respective repairs are hazard rate ordered. This is not the case for $\theta > 1$ (i.e. when the spare is worse than the original unit).

Another variation is to assume that X_i and X_i' are DID, that is, they are dependent and identically distributed. This is a reasonable assumption since they share the same environment. As in the preceding chapters, we can model this dependency through a survival copula $\widehat{C}$ which satisfies

$$\Pr(X_i > x, X_i' > y) = \widehat{C}(\bar{F}_i(x), \bar{F}_i(y))$$

for all x, y. Hence

$$\bar{F}_{Y_i}(t) = 2\bar{F}_i(t) - \widehat{C}(\bar{F}_i(t), \bar{F}_i(t)) = \bar{q}(\bar{F}_i(t)),$$

where $\bar{q}(u) = 2u - \widehat{C}(u, u)$ is a distortion function (which depends on $\hat{C}$) satisfying $\bar{q}(u) \geq u$ for all $u \in [0, 1]$ (since $\max(X_i, X_i') \geq X_i$).

Other interesting variations are to add $m - 1$ IID spares in parallel, which leads to the distortion function $\bar{q}_{m:m}(u) = 1 - (1 - u)^m \geq u$ (since $X_{m:m} \geq X_1$), or to add them with any other system structure with distortion $\bar{q}$ satisfying $\bar{q}(u) \geq u$ for all $u \in [0, 1]$.

Some cold redundancies can also be represented in this way (i.e. as distortions). If the lifetime of the resulting mechanism is $\widetilde{Y}_i = X_i + X_i'$, then its reliability is

$$\bar{F}_{\widetilde{Y}_i}(t) = \Pr(X_i + X_i' > t) = \bar{F}_i(t) + \int_0^t \Pr(X_i' > t - x | X_i = x) f_i(x) dx \quad (5.1)$$

for all $t \geq 0$, where $f_i = -\bar{F}_i'$ is the PDF of X_i. If X_i and X_i' are IID (perfect repair), then the reliability function of $\widehat{Y}_i = X_i + X_i'$ is

$$\bar{F}_{\widehat{Y}_i}(t) = \bar{F}_i(t) + \int_0^t \bar{F}_i(t - x) f_i(x) dx$$

which is the well know formula for the reliability function of a convolution. It is represented as $\bar{F}_{\widehat{Y}_i} = \bar{F}_i * \bar{F}_i$. In some models, this reliability can also be represented as a distortion (e.g. with exponential distributions). The same happen if they are dependent (see Navarro and Sarabia 2020).

However, if we consider a **minimal repair** (MR), that is,

$$\Pr(X_i' > y | X_i = x) = \frac{\bar{F}_i(x + y)}{\bar{F}_i(x)}$$

for all $x, y \geq 0$, then from (5.1), the reliability function of $\widetilde{Y}_i = X_i + X_i'$ is

$$\bar{F}_{\widetilde{Y}_i}(t) = \bar{F}_i(t) + \int_0^t \frac{\bar{F}_i(t)}{\bar{F}_i(x)} f_i(x) dx = \bar{F}_i(t) - \bar{F}_i(t) \log \bar{F}_i(t) = \bar{q}_{MR}(\bar{F}_i(t)) \quad (5.2)$$

for all $t \geq 0$, where

$$\bar{q}_{MR}(u) = u - u \log(u)$$

is a distortion function satisfying $\bar{q}_{MR}(u) \geq u$ for all $u \in [0, 1]$ (since $X_i + X_i' \geq X_i$). This model is also known as the **relevation transform** and it was introduced in formula (3.1) of Krakowski (1973) with the notation $\bar{F}_i \# \bar{F}_i$. In this model we can also consider some variations. For example we can consider m minimal repairs obtaining

$$\bar{q}_m(u) = u \sum_{i=0}^m \frac{1}{i!} (-\log(u))^i \quad (5.3)$$

with $\bar{q}_m(u) \geq u$ for all $u \in [0, 1]$ (since $X_i + X_i' + \cdots \geq X_i$).

We can also consider **imperfect repairs** (IR) with

$$\Pr(X_i' > y | X_i = x) = \frac{\bar{F}_i^\theta(x + y)}{\bar{F}_i^\theta(x)}$$

for all $x, y \geq 0$ and $\theta > 1$ (the spare is worse than the original component). This option leads to

$$\bar{F}_{Y_i}(t) = \bar{F}_i(t) + \int_0^t \frac{\bar{F}_i^\theta(t)}{\bar{F}_i^\theta(x)} f_i(x) dx = \bar{F}_i(t) - \frac{1}{1-\theta} \bar{F}_i^\theta(t) \left[1 - \bar{F}_i^{1-\theta}(t) \right] = \bar{q}_\theta^{IR}(\bar{F}_i(t))$$

for all $t \geq 0$, where

$$\bar{q}_\theta^{IR}(u) = \frac{\theta}{\theta - 1} u - \frac{1}{\theta - 1} u^\theta$$

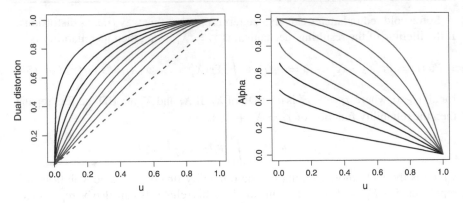

Fig. 5.4 Plots of $\bar{q}_\theta^{IR}$ (left) and α_θ^{IR} (right) for an imperfect repair with $\theta = 0.25, 0.5, 0.75$ (black), $\theta = 1$ (red, minimal repair) and $\theta = 1.5, 2, 3, 5$ (blue)

is a distortion function satisfying $\bar{q}_\theta^{IR}(u) \geq u$ for all $u \in [0, 1]$ (since $X_i + X_i' \geq X_i$). The case $0 < \theta < 1$ can be considered as well (although it could be unrealistic in some situations). Note that we obtain a negative mixture of $\bar{F}_i$ and $\bar{F}_i^\theta$. The plots of $\bar{q}_\theta^{IR}$ can be seen in Fig. 5.4, left. The case $\theta \to 1$ (red curve) represents the minimal repair case. As they are ordered, the respective repairs are ST ordered.

Its hazard rate can be written as

$$h_\theta^{IR}(t) = \alpha_\theta^{IR}(\bar{F}(t))h(t),$$

where $h = f/\bar{F}$ is the hazard rate of $\bar{F}$ and

$$\alpha_\theta^{IR}(u) = \frac{u(\bar{q}_\theta^{IR})'(u)}{\bar{q}_\theta^{IR}(u)} = \theta \frac{1 - u^{\theta-1}}{\theta - u^{\theta-1}}$$

for $u \in [0, 1]$. The plots of α_θ^{IR} can be seen in Fig. 5.4, right. As they are ordered, the respective repairs are hazard rate ordered.

We conclude this section by comparing the three main replacement options. Of course, if $Y_i = \max(X_i, X_i')$ (hot spare parallel), then $Y_i \leq X_i + X_i'$ and so, in particular, when they are independent $\bar{F}_{Y_i} \leq \bar{F}_i * \bar{F}_i$ (perfect repair or convolution). Under minimal repair $\bar{F}_{Y_i} \leq \bar{F}_i \# \bar{F}_i$ holds since

$$\bar{q}_{2:2}(u) = 2u - u^2 \leq \bar{q}_{MR}(u) = u - u \log(u)$$

for all $u \in [0, 1]$. Even more, as $\bar{q}'_{MR}/\bar{q}'_{2:2}$ is decreasing, then $Y_i \leq_{LR} \tilde{Y}_i$ for all F, where $\tilde{Y}_i$ represents the total lifetime from the beginning under a minimal repair. To compare $\tilde{Y}_i$ (minimal repair) and $\hat{Y}_i = X_i + X_i'$ (perfect repair or convolution) when X_i and X_i' are IID we have the following result.

Proposition 5.1 *With the notation introduced above, if F_i is NBU (NWU), then* $\tilde{Y}_i \leq_{ST} \hat{Y}_i (\geq_{ST})$.

Proof Recall that NBU means that $\bar{F}_i(x)\bar{F}_i(y) \geq \bar{F}_i(x+y)$ for all $x, y \geq 0$. Hence, from (5.2), we get

$$\bar{F}_{\widetilde{Y}_i}(t) = \bar{F}_i(t) + \int_0^t \frac{\bar{F}_i(t)}{\bar{F}_i(x)} f_i(x)dx \leq \bar{F}_i(t) + \int_0^t \bar{F}_i(t-x)f_i(x)dx = \bar{F}_{\widehat{Y}_i}(t)$$

since $\bar{F}_i(t-x)\bar{F}_i(x) \geq \bar{F}_i(t)$ for all $0 \leq x \leq t$. The inequality is reversed for NWU distributions.   □

Note that, for the "natural" aging property (NBU), the perfect repair is better than the minimal repair (as expected) and we can write

$$Y_i \leq_{ST} \widetilde{Y}_i \leq_{ST} \widehat{Y}_i.$$

For the dual class (NWU) we get

$$Y_i \leq_{ST} \widehat{Y}_i \leq_{ST} \widetilde{Y}_i.$$

Of course, for the exponential distribution (which is both NBU and NWU), we have $\widetilde{Y}_i =_{ST} \widehat{Y}_i$, that is, minimal and perfect repairs coincide.

5.2 Systems with ID Components

In this case we can compare different repair policies by using the ordering results for distorted distributions obtained in Chap. 3. Recall that, in the general case, the system reliability can be written as

$$\bar{F}_T(t) = \bar{Q}(\bar{F}_1(t), \ldots, \bar{F}_n(t)),$$

where $\bar{Q}$ is a distortion function. If we assume that the components are ID, that is, $\bar{F}_1 = \cdots = \bar{F}_n = \bar{F}$ say, then this representation can be reduced to

$$\bar{F}_T(t) = \bar{Q}(\bar{F}(t), \ldots, \bar{F}(t)) = \bar{q}(\bar{F}(t)),$$

where $\bar{q}(u) = \bar{Q}(u, \ldots, u)$ is a distortion function.

If we apply a redundancy policy $\mathbf{r} = (r_1, \ldots, r_n)$ where the redundancy applied to the ith components is represented by $\bar{q}_{r_i}$, then the reliability function of the lifetime $T_{\mathbf{r}}$ of the resulting system can be written as

$$\bar{F}_{\mathbf{r}}(t) = \bar{Q}(\bar{q}_{r_1}(\bar{F}(t)), \ldots, \bar{q}_{r_n}(\bar{F}(t))) = \bar{q}_{\mathbf{r}}(\bar{F}(t)),$$

where

$$\bar{q}_{\mathbf{r}}(u) = \bar{Q}(\bar{q}_{r_1}(u), \ldots, \bar{q}_{r_n}(u))$$

for $u \in [0, 1]$. Note that if we do not apply redundancy to the ith component, then $\bar{q}_{r_i}(u) = u$. Of course, we always get $T \leq_{ST} T_{\mathbf{r}}$ since $\bar{Q}$ is increasing and we assume $\bar{q}_{r_i}(u) \geq u$ for $i = 1, \ldots, n$.

If we have another redundancy policy $\mathbf{s} = (s_1, \ldots, s_n)$, then the reliability function of the resulting system can be represented in a similar way with another distortion function $\bar{q}_{\mathbf{s}}$. Hence $T_{\mathbf{r}}$ and $T_{\mathbf{s}}$ can be compared just by comparing their respective distortion functions using Proposition 3.2.

The comparisons under minimal repairs were studied in Arriaza et al. (2018). Here $\mathbf{r} = (r_1, \ldots, r_n)$ means that r_i minimal repairs are applied to the ith component, with $r_i \geq 0$ for $i = 1, \ldots, n$. With this notation we can obtain the following classic result for series systems with IID components that can be traced back to Shaked and Shanthikumar (1992), Result 2.4(s) (see also Theorem 4 in Li and Ding 2010). Obviously, in the case of series systems with IID components, the repair strategy given by the vector $\mathbf{r} = (r_1, \ldots, r_n)$ is the same as that of $\mathbf{r}' = (r_{\pi(1)}, \ldots, r_{\pi(n)})$ for any permutation $\pi : \{1, \ldots, n\} \to \{1, \ldots, n\}$. So, without loss of generality, we can assume for this system that $r_1 \geq \ldots \geq r_n$. Moreover, we have $\bar{Q}(u_1, \ldots, u_n) = u_1 \ldots u_n$ and so

$$\bar{q}_{\mathbf{r}}(u) = \bar{q}_{r_1}(u) \ldots \bar{q}_{r_n}(u)$$

for $u \in [0, 1]$, where these distortions functions are defined as in (5.3). Hence we have the following theorem.

Theorem 5.1 (Shaked and Shanthikumar 1992) *Consider a series system with n IID components with a common reliability function $\bar{F}$. Suppose that we have available $m \in \mathbb{Z}_+$ minimal repairs that can be freely allocated to any component. Let $p, s \in \mathbb{Z}_+$ be the unique integer numbers such that $m = pn + s$ and $0 \leq s < n$. Then, the optimal allocation strategy, in terms of the usual stochastic order, is given by the vector*

$$\mathbf{r}^\star = (\overbrace{p+1, \; p+1, \ldots, \; p+1}^{s}, \overbrace{p, \; p, \ldots, \; p}^{n-s}).$$

As expected, the best option is to distribute all the available repairs as much as possible between the components. An alternative proof to that given in Shaked and Shanthikumar (1992) is provided in Arriaza et al. (2018). It is interesting to note here that if the optimal allocation strategy cannot be applied due to some other external constraint, then using the sequence $\{\mathbf{r}_i\}_{i \in \{1, \ldots, \upsilon\}}$ defined in this proof we always have available the second best choice as optimal strategy, and so on (or a path to improve the initial strategy $\mathbf{r}$). Also note that as a consequence of the proof, the worst option is always $(m, 0, \ldots, 0)$, i.e., to assign all the repairs to a fixed component.

We can also compare repairs in any other system structures. Let us see an example extracted from Arriaza et al. (2018). Additional results for minimal repairs can be seen in Navarro et al. (2019). Similar results can be obtained for other repair options based on distortions.

Example 5.1 Consider a 2-out-of-3 system with IID components with a common reliability function $\bar{F}$. Assume a fixed number $m = 7$ of available minimal repairs. Let us study all the possible ST comparisons of lifetimes $T_{\mathbf{r}}$ obtained from the repair policies $\mathbf{r} = (r_1, r_2, r_3) \in \mathbb{Z}_+^3$ with $r_1 \geq r_2 \geq r_3$ and $r_1 + r_2 + r_3 = 7$. Note that in this case they are also equivalent under permutations in $\mathbf{r}$. Firstly, given $\mathbf{r} = (r_1, r_2, r_3) \in \mathbb{Z}_+^3$ and assuming that the component lifetimes are independent, we obtain that the reliability function of the system lifetime $T_{\mathbf{r}}$ associated to $\mathbf{r}$ is

$$\bar{F}_{\mathbf{r}}(t) = \bar{Q}(\bar{F}_{(r_1)}(t), \bar{F}_{(r_2)}(t), \bar{F}_{(r_3)}(t)) = \bar{q}_{\mathbf{r}}(\bar{F}(t)),$$

where $\bar{Q}(u, v, w) = uv + uw + vw - 2uvw$,

$$\bar{q}_{\mathbf{r}}(u) = \bar{q}_{r_1}(u)\bar{q}_{r_2}(u) + \bar{q}_{r_1}(u)\bar{q}_{r_3}(u) + \bar{q}_{r_2}(u)\bar{q}_{r_3}(u) - 2\bar{q}_{r_1}(u)\bar{q}_{r_2}(u)\bar{q}_{r_3}(u)$$

and $\bar{q}_{r_i}$ is the distortion function given in (5.3) for $i = 1, 2, 3$. Then, we have that

$$T_{\mathbf{r}_1} \leq_{st} T_{\mathbf{r}_2} \text{ for all } \bar{F} \Leftrightarrow \bar{q}_{\mathbf{r}_1}(u) \leq \bar{q}_{\mathbf{r}_2}(u) \text{ for all } u \in (0, 1).$$

Therefore, if we want to compare two strategies $\mathbf{r}_1$ and $\mathbf{r}_2$, we just need to plot both functions, $\bar{q}_{\mathbf{r}_1}$ and $\bar{q}_{\mathbf{r}_2}$ on the interval $[0, 1]$. For instance, in this way we can confirm that $T_{\mathbf{r}_1} \leq_{ST} T_{\mathbf{r}_2}$ for all reliability functions $\bar{F}$ when $\mathbf{r}_1 = (7, 0, 0)$ and $\mathbf{r}_2 = (6, 1, 0)$. We will write $\mathbf{r}_1 \to \mathbf{r}_2$ to denote that the strategy $\mathbf{r}_1$ is better than $\mathbf{r}_2$ or, in other words, $T_{\mathbf{r}_2} \leq_* T_{\mathbf{r}_1}$ holds for a given order $\leq_*$.

Following the previous procedure, we obtain the graphs given in Fig. 5.5 with all the relationships for the comparisons of the repair strategies in the HR order (left) and in the ST order (right). The strategies that are not connected in the graph represent lifetimes of systems that are not comparable in the usual stochastic order (respectively, in the hazard rate order). In this case an optimal allocation strategy does not exist in terms of the usual stochastic order. Note that, a priori, all the minimal path sets of the 2-out-of-3 system are equally important due to the structure of the system. Note that the replacement policy represented by the vector $\mathbf{r}^\star = (4, 3, 0)$ (which applies all the repairs to the components in the first path set) is ordered with a larger number of alternatives (see Fig. 5.5). However, $\mathbf{r}^\star$ is not stochastically ordered neither with $(3, 3, 1)$ nor with $(3, 2, 2)$. Similar comments hold for the HR order.

5.3 Systems with Non-ID Components

As in the preceding section, we know that the reliability function of the system lifetime T can be written as

$$\bar{F}_T(t) = \bar{Q}(\bar{F}_1(t), \ldots, \bar{F}_n(t))$$

for all t, where $\bar{Q}$ is a distortion function. Hence, if we apply a redundancy with distortion $\bar{q}(u) \geq u$ to the ith component, the reliability function of the resulting system lifetime T_i is

$$\bar{F}_{T_i}(t) = \bar{Q}_i(\bar{F}_1(t), \ldots, \bar{F}_n(t))$$

with

$$\bar{Q}_i(u_1, \ldots, u_n) = \bar{Q}(u_1, \ldots, u_{i-1}, \bar{q}(u_i), u_{i+1} \ldots, u_n)$$

for $0 \leq u_i \leq 1$ and $i = 1, \ldots, n$. Note that we are assuming a common redundancy mechanism (distortion) for all the components.

Of course, then we have $T \leq_{ST} T_i$ for $i = 1, \ldots, n$. However, we want to compare T_i and T_j to determine where the redundant component should be placed.

In the first result, extracted from Navarro and Fernández-Martínez (2021), we analyze series systems with independent components. In this case, we just compare T_1 and T_2.

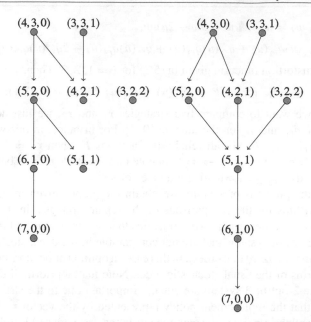

Fig. 5.5 Relationships among all possible lifetimes T_r after seven minimal repairs by using the hazard rate order (left) and the usual stochastic order (right) for the 2-out-of-3 system considered in Example 5.1

Proposition 5.2 *Let* $T = \min(X_1, \ldots, X_n)$ *with independent components.*

(i)   *If* $X_1 \geq_{ST} X_2$ *and*

$$\frac{\bar{q}(u)}{u} \text{ is decreasing in } (0, 1), \tag{5.4}$$

   then $T_1 \leq_{ST} T_2$ *for all* $F_3, \ldots, F_n$.

(ii) *If* $X_1 \geq_{HR} X_2$ *and*

$$\frac{\bar{q}(uv)}{v\bar{q}(u)} \text{ is decreasing in } (0, 1)^2, \tag{5.5}$$

   then $T_1 \leq_{HR} T_2$ *for all* $F_3, \ldots, F_n$.

(iii) *The condition (5.4) holds iff* $T \leq_{HR} T_i$ *for all* $F_1, \ldots, F_n$ *and* $i = 1, \ldots, n$.

Proof (i) The condition $T_1 \leq_{ST} T_2$ holds iff

$$\bar{Q}_1(u_1, \ldots, u_n) = \bar{q}(u_1)u_2 \ldots u_n \leq \bar{u}_1 q(u_2) \ldots u_n = \bar{Q}_2(u_1, \ldots, u_n)$$

which is equivalent to

$$\bar{q}(u_1)u_2 \leq \bar{u}_1 q(u_2).$$

As we assume $\bar{F}_1 \geq \bar{F}_2$ and (5.4), we get

$$\frac{\bar{q}(\bar{F}_1(t))}{\bar{F}_1(t)} \leq \frac{\bar{q}(\bar{F}_2(t))}{\bar{F}_2(t)}$$

and so $T_1 \leq_{ST} T_2$ for all $F_3, \ldots, F_n$.

(ii) The condition $T_1 \leq_{HR} T_2$ holds if and only if

$$\frac{\bar{Q}_2(\bar{F}_1(t), \ldots, \bar{F}_n(t))}{\bar{Q}_1(\bar{F}_1(t), \ldots, \bar{F}_n(t))} \text{ is increasing in } t,$$

which is equivalent to

$$\frac{\bar{F}_1(t)\bar{q}(\bar{F}_2(t))}{\bar{F}_2(t)\bar{q}(\bar{F}_1(t))} \text{ is increasing in } t$$

As we assume $X_1 \geq_{HR} X_2$, $g(t) = \bar{F}_2(t)/\bar{F}_1(t)$ is decreasing in t. Hence $g(t) \in [0, 1]$. Then, by applying (5.5) to $u = \bar{F}_1(t)$ and $v = g(t)$, we get that

$$\frac{\bar{F}_1(t)\bar{q}(\bar{F}_2(t))}{\bar{F}_2(t)\bar{q}(\bar{F}_1(t))}$$

is increasing in t and so $T_1 \leq_{HR} T_2$ holds for all $F_3, \ldots, F_n$.

(iii) The condition $T \leq_{HR} T_i$ holds if and only if

$$\frac{\bar{Q}_i(\bar{F}_1(t), \ldots, \bar{F}_n(t))}{\bar{Q}(\bar{F}_1(t), \ldots, \bar{F}_n(t))} \text{ is increasing in } t,$$

which is equivalent to

$$\frac{\bar{q}(\bar{F}_i(t))}{\bar{F}_i(t)} \text{ is increasing in } t$$

for all $\bar{F}_i$. This property is equivalent to (5.4).   □

Note that (i) means that, under condition (5.4), the redundant component should be applied to the strongest components (in the ST order). To extend this property to the HR order we need the stronger condition (5.5). The meaning of (5.4) can be seen in (iii). It is equivalent to the condition: T_i is HR better than T and to the same ordering property for the original component X_i and the resulting redundancy mechanism Y_i.

The property (5.4) is satisfied for the usual redundancy mechanism. For example, for a hot IID spare added in parallel we have

$$\frac{\bar{q}_{2:2}(u)}{u} = \frac{2u - u^2}{u} = 2 - u$$

which is decreasing. The same happen for m independent spares added in parallel.

For a cold standby unit with minimal repair we have

$$\frac{\bar{q}_{MR}(u)}{u} = \frac{u - u \log u}{u} = 1 - \log u$$

that is also decreasing. Hence (5.4) holds. The same happen for m minimal repairs.

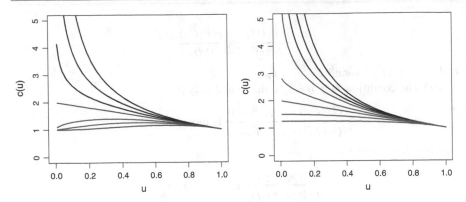

Fig. 5.6 Plots of $c(u) = \bar{q}_\theta(u)/u$ for a hot spare in parallel with reliability $\bar{F}^\theta$ (left) and an imperfect repair (right) with $\theta = 0.25, 0.5, 0.75$ (black), $\theta = 1$ (red, minimal repair) and $\theta = 1.5, 2, 3, 5$ (blue)

However, (5.4) is not always true. For example, if we add a spare in parallel with reliability $\bar{F}^\theta$, we obtain the plots in Fig. 5.6 (left) for $c(u) = \bar{q}_\theta(u)/u$. There we can see that function c is decreasing for $0 < \theta \leq 1$ but that it is not monotone for $\theta > 1$ (since $c(0) = c(1) = 1$).

However, (5.4) holds for imperfect repairs since

$$c(u) = \frac{\bar{q}_\theta(u)}{u} = \frac{\theta - u^{\theta-1}}{\theta - 1}$$

is decreasing in u for all $\theta > 0$ (see Fig. 5.6, right). As mentioned above, it is also decreasing for a minimal repair (red curve).

The condition (5.5) is not so common. For example, it fails in active redundancies since

$$\frac{\bar{q}(uv)}{v\bar{q}(u)} = \frac{2uv - u^2v^2}{2uv - u^2v} = \frac{2 - uv}{2 - u}$$

is increasing in u and decreasing in v in the set $(0, 1)^2$. The same happen for minimal repairs since

$$\frac{\bar{q}(uv)}{v\bar{q}(u)} = \frac{uv - uv\log(uv)}{uv - uv\log(u)} = 1 + \frac{-\log(v)}{1 - \log(u)}$$

is increasing in u and decreasing in v in the set $(0, 1)^2$.

Similar (reverse) results can be obtained for parallel systems with independent components. For example, if $X_1 \geq_{ST} X_2$ and $q(u)/u$ is increasing, then $T_1 \geq_{ST} T_2$ for all $F_3, \ldots, F_n$, that is, in this system, it is better to reinforce the strongest component (as expected). For other system structures the answer is not so clear, see Navarro and Fernández-Martínez (2021). The same happen if we consider dependent components. In these cases they can be compared by using distortions.

We conclude this section by establishing comparisons between redundancies at components' or system's levels. The BP (Barlow and Proschan) principle for active redundancies in parallel is established in the following theorem. It was given in Theorem 2.4 of Barlow and Proschan (1975), p. 8, (see also Samaniego 2007, p. 17).

Theorem 5.2 (BP-principle) *If we consider active redundancies added in parallel and the component and spares lifetimes have the same joint distribution, then the system with redundancy at components' level is always ST better than the system with redundancy at system's level.*

Proof We provide the proof for an active redundancy. The proof for m active redundancies is similar. If $X_1, \ldots, X_n$ are the components' lifetimes and $X_1', \ldots, X_n'$ are the spares' lifetimes. We assume that in both redundancy options $(X_1, \ldots, X_n, X_1', \ldots, X_n')$ has the same joint distribution or, equivalently, that both systems are built with the same components.

Let $P_1, \ldots, P_r$ be the minimal path sets of the original system. Then the minimal path sets of the system with redundancy at system's level are $P_1, \ldots, P_r, P_1', \ldots, P_r'$, where P_i' is the set with the spares of the components in the set P_i. It is easy to see that all these sets are also path sets of the system with redundancy at components' level. Hence, if we assume that they have the same components, the system with redundancy at components' level works whenever the system with redundancy at system's level does so. Hence, their lifetimes are ordered for sure and so we have the ST order when the components have the same joint distribution (see Theorem 1.A.1 in Shaked and Shanthikumar 2007, p. 5). □

We must say that the assumption about a common joint distribution for the components and spares is quite unrealistic when the components are dependent (since the spares are placed at different positions). However, it holds when the components and spares are independent. Let us see an example.

Example 5.2 Let us consider the system with lifetime

$$T = \max(X_1, \min(X_2, X_3))$$

and independent components. Its dual distortion function is

$$\bar{Q}(u_1, u_2, u_3) = u_1 + u_2 u_3 - u_1 u_2 u_3$$

for $u_1, u_2, u_3 \in [0, 1]$.

The lifetime of the system with redundancy at the components' level is

$$T_1 = \max(\max(X_1, X_1'), \min(\max(X_2, X_2'), \max(X_3, X_3'))),$$

where X_1', X_2', X_3' represent the lifetimes of the spares. If we assume that the components and the spares are independent and that $X_i =_{ST} X_i'$ for $i = 1, 2, 3$, then the dual distortion function of T_1 is

$$\bar{Q}_1(u_1, u_2, u_3) = \bar{q}_{2:2}(u_1) + \bar{q}_{2:2}(u_2)\bar{q}_{2:2}(u_3) - \bar{q}_{2:2}(u_1)\bar{q}_{2:2}(u_2)\bar{q}_{2:2}(u_3)$$

for $u_1, u_2, u_3 \in [0, 1]$, where $\bar{q}_{2:2}(u) = 2u - u^2$ for $u \in [0, 1]$.

Analogously, the lifetime of the system with redundancy at the system's level is

$$T_2 = \max(\max(X_1, \min(X_2, X_3)), \max(X_1', \min(X_2', X_3')))$$

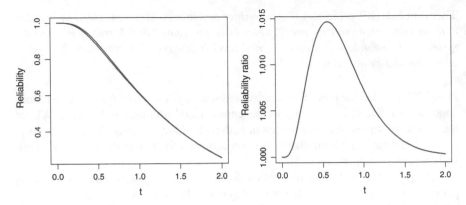

Fig. 5.7 Plots of the reliability functions (left) and its ratio (right) for the systems in Example 5.2 with redundancies at components' level (blue) and at system's level (red)

that is, in this case we have two independent copies of the system connected in parallel. If we assume the same joint distribution for components and spares as in the preceding case, then its dual distortion function is

$$\bar{Q}_2(u_1, u_2, u_3) = \bar{q}_{2:2}(\bar{Q}(u_1, u_2, u_3))$$

for $u_1, u_2, u_3 \in [0, 1]$.

Hence, from the preceding theorem (BP-principle), we have $T_1 \geq_{ST} T_2$ for all F_1, F_2, F_3. The respective reliability functions can be seen in Fig. 5.7 (left) for exponential components with hazard rates 1, 2, 3, respectively. Note that the reliabilities are very similar. The ratio in the right plot shows that this property cannot be extended to the hazard rate order.

5.4 Importance Indices

There exist several importance indices for the components in a system, especially in the case of independent components, see for example Barlow and Proschan (1975) and Kuo and Zhu (2012). Some of them only depend on the structure of the system, while others also depend on the components' distributions.

For example, the structural importance of the ith component is defined (see Barlow and Proschan 1975, p. 13) as

$$n_\phi(i) = \frac{1}{2^{n-1}} \sum_{x_j=0,1, j \neq i} [\phi(x_1, \ldots, 1, \ldots, x_n) - \phi(x_1, \ldots, 0, \ldots, x_n)],$$

where the ones and zeros are placed at the ith positions. This measure takes into account how many times the ith component is crucial for the system. If we consider a 2-out-of-3 system, then $n_\phi(i) = 1/2$ for $i = 1, 2, 3$ while if $\phi(x_1, x_2, x_3) = \min(x_1, \max(x_2, x_3))$, then $n_\phi(1) = 3/4$ and $n_\phi(i) = 1/4$ for $i = 2, 3$. The main advantage is that these indices can always be compared

Another popular index is the Barlow and Proschan (BP) importance measure defined as

$$m(i) = \Pr(T = X_i).$$

This index depends on the components' distributions. If we assume IID components with a common continuous distribution (no ties), then this index only depends on the structure and so it can be written as $m_\phi(i)$. For example, in a 2-out-of-3 system, $m_\phi(i) = 1/3$ for $i = 1, 2, 3$ while if $\psi(x_1, x_2, x_3) = \min(x_1, \max(x_2, x_3))$, then $m_\phi(1) = 4/6$ and $m_\phi(i) = 1/6$ for $i = 2, 3$. Note that in this index, with no ties, we have $\sum_{i=1}^n m(i) = 1$.

In the case of independent components, another popular index based on the reliability function of the structure $\bar{Q}_\Pi$ is

$$I_\phi(i) = \partial_i \bar{Q}_\Pi(u_1, \ldots, u_n),$$

where remember that $\bar{Q}_\Pi$ is also the dual distortion function based on the product copula (or the function obtained with the pivotal decomposition). It is known as the Birnbaum (B) importance measure (see Birnbaum 1969) and it can also be written as

$$I_\phi(i) = \bar{Q}(u_1, \ldots, u_{i-1}, 1, u_{i+1}, \ldots, u_n) - \bar{Q}(u_1, \ldots, u_{i-1}, 0, u_{i+1}, \ldots, u_n)$$

or as

$$I_\phi(i) = E(\phi(X_1, \ldots, 1, \ldots, X_n) - \phi(X_1, \ldots, 0, \ldots, X_n)),$$

where $X_1, \ldots, X_n$ are IID with $\Pr(X_i = 1) = u_i$ and $\Pr(X_i = 0) = 1 - u_i$ for $i = 1, \ldots, n$ (see Barlow and Proschan 1975, p. 22). The main disadvantage is that this index is not a number but a function of $u_1, \ldots, u_n$. So the indices for the different components cannot be compared.

For example, for a 2-out-of-3 system, we get

$$I_\phi(1) = u_2 + u_3 - 2u_2u_3,$$
$$I_\phi(2) = u_1 + u_3 - 2u_1u_3$$

and

$$I_\phi(3) = u_1 + u_2 - 2u_1u_2.$$

However, if we consider the system $\phi(x_1, x_2, x_3) = \min(x_1, \max(x_2, x_3))$, then

$$I_\phi(1) = u_2 + u_3 - 2u_2u_3,$$
$$I_\phi(2) = u_1 - u_1u_3$$

and

$$I_\phi(3) = u_1 - u_1u_2.$$

If the components are IID, we can assume $u = u_1 = u_2 = u_3$ and then

$$I_\phi(i) = 2u - 2u^2, \ i = 1, 2, 3$$

in the 2-out-of-3 system (i.e. all the components have the same importance) while

$$I_\phi(1) = 2u - u^2 \geq u - u^2 = I_\phi(i), \ i = 2, 3$$

in the other system (the first component is more important than the others). These are expectable properties.

If we consider dependent components, the index should also take into account the dependence structure. However, it should not depend on the components' distributions. It should also be used to determine the best replacement positions. Actually, we may want to place the best components or the redundancies at the most critical (important) positions.

In this case (dependent components) the three equivalent expressions considered above for $I_\phi(i)$ lead to different options. The most useful one in practice is

$$I_{\phi,C}(i) = \partial_i \bar{Q}_C(u_1, \ldots, u_n), \tag{5.6}$$

where $\bar{Q}_C$ is the dual distortion function of the system when the dependence is determined by the copula C. Note that it also depends on the system structure ϕ. To simplify the notation we will just write I_i instead of $I_{\phi,C}(i)$. The meaning is clear the most important components are those which lead to a higher increment in the system reliability function (when they are improved).

Again the indices are functions of $u_1, \ldots, u_n$. However, as above, we could consider ID components and then they are just functions of $u = u_1 = \cdots = u_n$ with $I_i(u) := I_i(u, \ldots, u)$.

This index was analyzed in Miziuła and Navarro (2019) for dependent components proving that

$$m(i) = \Pr(T = X_i) = \int_0^\infty I_i(\bar{F}_1(t), \ldots, \bar{F}_n(t)) \, dF_i(t)$$

for $i = 1, \ldots, n$. In particular, if the components are ID, then

$$m(i) = \Pr(T = X_i) = \int_0^1 I_i(u) \, du. \tag{5.7}$$

In this case, $\Pr(T = X_i)$ does not depend on $F = F_1 = \cdots = F_n$ and, if $I_i(u) \leq I_j(u)$ for all $u \in [0, 1]$, then $m(i) \leq m(j)$ for all F.

This index can also be used to determine the best replacement position. The result extracted from Theorem 2.4 of Navarro et al. (2020) can be stated as follows. Its proof can be seen there.

Theorem 5.3 *If* $I_1(u_1, \ldots, u_n) \leq I_2(u_1, \ldots, u_n)$ *for all* $u_1, \ldots, u_n$, *then* $T_1 \leq_{ST} T_2$, *where* T_i *is the system obtained by applying a redundancy with dual distortion* $\bar{q}$ *to the* ith *component for* $i = 1, 2$.

The good point of the preceding theorem is that it holds for arbitrary redundancies satisfying $\bar{q}(u) \geq u$ for $u \in [0, 1]$. It can also be applied to mixed systems. However, the condition $I_1 \leq I_2$ assumed there is too strong. So some weaker conditions that lead to the similar result were analyzed in Navarro et al. (2020). Other conditions for

specific redundancies (active redundancy in parallel or minimal repair) are analyzed
as well.

Problems

1. Prove equation (5.3).
2. Compare two repair policies in a system with ID components.
3. Check an arrow in Fig. 5.5.
4. Study if Theorem 5.1 can be extended to hot redundancies of independent components added in parallel (Indication: Try to prove it first for $n = 2$).
5. Study the redundancy policies considered in Example 5.1 but using hot independent spares connected in parallel.
6. Compare a redundancy at different positions in a system with IID components.
7. Compare a redundancy at different positions in a system with independent components.
8. Compare a redundancy at different positions in a system with dependent components.
9. Confirm the BP-principle in a system with independent components.
10. Compute the BP and B importance measures in a system with IID components and confirm that (5.7) holds.
11. Compute the BP and B importance measures in a system with DID components and confirm that (5.7) holds.

Glossary

A^c	complementary (contrary) of the set (event) A
$\|A\|$	cardinality of the set A
$(a_{1:n}, \ldots, a_{n:n})$	the increasing ordered vector obtained from $\mathbf{a} = (a_1, \ldots, a_n)$
$\mathbf{a} \leq_m \mathbf{b}$	majorization order: $\sum_{i=1}^{j} a_{i:n} \leq \sum_{i=1}^{j} b_{i:n}$ for $j = 1, 2, \ldots, n - 1$ and $\sum_{i=1}^{n} a_{i:n} = \sum_{i=1}^{n} b_{i:n}$
$\mathbf{0}_i(\mathbf{x})$	$(x_1, \ldots, x_{i-1}, 0, x_{i+1}, \ldots, x_n)$
$\mathbf{1}_i(\mathbf{x})$	$(x_1, \ldots, x_{i-1}, 1, x_{i+1}, \ldots, x_n)$
$\mathbf{1}_J$	$(x_1, \ldots, x_n)$ with $x_i = 1$ if $i \in J$ and $x_i = 0$ if $i \notin J$
1_A	indicator Boolean function, $1_A = 1$ (resp. 0) if A is true (false)
$2^{[n]}$	set with all the subsets of $[n]$
$[n]$	set $\{1, \ldots, n\}$ for $n = 1, 2, \ldots$
$\mathbf{t}_P$	$(z_1, \ldots, z_n)$ with $z_i = t$ for $i \in P$ and $z_i = -\infty$ for $i \notin P$
$\mathbf{t}^P$	$(z_1, \ldots, z_n)$ with $z_i = t$ for $i \in P$ and $z_i = \infty$ for $i \notin P$
$\mathbf{t}_k$	$(z_1, \ldots, z_n)$ with $z_i = t$ for $i = 1, 2, \ldots, k$ and $z_i = -\infty$ for $i = k+1, k+2, \ldots, n$
$\mathbf{t}^k$	$(z_1, \ldots, z_n)$ with $z_i = t$ for $i = 1, 2, \ldots, k$ and $z_i = \infty$ for $i = k+1, k+2, \ldots, n$
$X_{i:n}$	ith order statistic from $X_1, \ldots, X_n$ for $i = 1, \ldots, n$
$X_{1:n}$	$\min(X_1, \ldots, X_n)$
$X_{n:n}$	$\max(X_1, \ldots, X_n)$
X_P	$\min_{i \in P} X_i$
X^P	$\max_{i \in P} X_i$

© The Editor(s) (if applicable) and The Author(s), under exclusive license
to Springer Nature Switzerland AG 2022
J. Navarro, *Introduction to System Reliability Theory*,
https://doi.org/10.1007/978-3-030-86953-3

References

Abouammoh, A. and El-Neweihi, E. (1986). Closure of NBUE and DMRL under the formation of parallel systems. *Statistics and Probability Letters* **4**, 223–225.

Arcones, M.A., Kvam, P.H. and Samaniego, F.J. (2002). Nonparametric estimation of a distribution subject to a stochastic precedence constraint. *Journal of the American Statistical Association* **97**, 170–182

Arnold, B.C., Balakrishnan, N. and Nagaraja, H.N. (2008). *A First Course in Order Statistics*. SIAM.

Arriaza, A., Navarro, J. and Suárez-Llorens, A. (2018). Stochastic comparisons of replacement policies in coherent systems under minimal repair. *Naval Research Logistics* **65**, 550–565.

Aven, T. and Jensen, U. (1999). Stochastic Models in Reliability. Second Edition. Springer, New York.

Barlow, R.E. and Proschan, F. (1975). *Statistical Theory of Reliability and Life Testing*. Holt, Rinehart and Winston, New York.

Belzunce, F., Martínez-Riquelme, C. and Mulero, J. (2016). *An Introduction to Stochastic Orders*. Elsevier, London.

Belzunce, F., Martínez-Riquelme, C. and Ruiz J.M. (2013). On sufficient conditions for mean residual life and related orders. *Computational Statistics and Data Analysis* **61**, 199–210.

Birnbaum, Z. (1969). On the importance of different components in a multicomponent system. In *Multivariate Analysis*, P. Krishnaiah, Ed., New York, Academic, pp. 581–592.

Block, H.W. and Savits, T.H. (1982). A decomposition for multistate monotone systems. *Journal of Applied Probability* **19**, 391–402.

Block, H.W., Savits, T.H., and Shaked, M. (1982). Some concepts of negative dependence. *Annals of Probability* **10**, 765–772.

Block, H.W., Savits, T.H. and Singh H. (1998). The reversed hazard rate function. *Probability in the Engineering and Informational Sciences* **12**, 69–90.

Boland, P. (2001). Signatures of indirect majority systems. *Journal of Applied Probability* **38**, 597–603.

Boland, P.J. and Samaniego, F. (2004). The signature of a coherent system and its applications in Reliability. In: *Mathematical Reliability: An Expository Perspective*. Edited by R. Soyer, T. Mazzuchi and N. Singpurwalla, vol. 67 in the International Series in Operational Research and Management Science, Kluwer, 1–29.

J. Navarro, *Introduction to System Reliability Theory*,
https://doi.org/10.1007/978-3-030-86953-3

Borgonovo, E. (2010). The reliability importance of components and prime implicants in coherent and non-coherent systems including total-order interactions. European Journal of Operational Research **204**, 485–495.

Bryson, M.C. and Siddiqui, M.M. (1969). Some criteria for aging. *Journal of the American Statistical Association* **64**, 1472–1483.

Burkschat, M. and Navarro, J. (2018). Stochastic comparisons of systems based on sequential order statistics via properties of distorted distributions. *Probability in the Engineering and Informational Sciences* **32**, 246–274.

Cha, J.H. and Finkelstein, M. (2018). Point Processes for Reliability Analysis. Shocks and Repairable Systems. Springer.

Coolen, F.P.A. and Coolen-Maturi, T. (2012). On generalizing the signature to systems with multiple types of components. In: *Complex Systems and Dependability*. Edited by Zamojski et al., Springer, New York, 115–130.

David, H.A. and Nagaraja, H.N. (2003). *Order Statistics*, Third edition. Wiley, Hoboken, New Jersey.

Di Crescenzo, A. (2007). A Parrondo paradox in reliability theory. *The Mathematical Scientist* **32**, 17–22.

Durante, F. and Papini, P.L. (2007). A weakening of Schur-concavity for copulas. *Fuzzy Sets and Systems* **158**, 1378–1383.

Durante, F. and Sempi, C. (2016). *Principles of Copula Theory*. CRC/Chapman & Hall, London.

Esary, J.D., Marshall, A.W. and Proschan, F. (1970). Some reliability applications of the hazard transform. *SIAM Journal on Applied Mathematics* **18**, 849–860.

Esary, J. and Proschan, F. (1963). Relationship between system failure rate and component failure rates. *Technometrics* **5**, 183–189.

Fantozzi, F., and Spizzichino, F. (2015). Multi-attribute target-based utilities and extensions of fuzzy measures. *Fuzzy Sets and Systems* **259**, 29–43.

Gertsbakh, I.B. and Shpungin, Y. (2010). *Models of network reliability. Analysis, combinatorics, and Monte Carlo.* CRC Press, Boca Raton, FL.

Gertsbakh, I. and Shpungin, Y. (2020). *Network Reliability. A Lecture Course.* Springer Briefs in Electrical and Computer Engineering. Springer, Singapore.

Glaser, R.E. (1980). Bathtub and related failure rate characterizations. *Journal of the American Statistical Association* **75**, 667–672.

Grabisch, M. (2016). *Set, Functions, Games and Capacities in Decision Making.* Springer.

Grabisch, M., Marichal, J.-L., Mesiar, R. and Pap, E. (2009). *Aggregation Functions.* Cambridge University Press, Cambridge, UK.

Hürlimann, W. (2004). Distortion risk measures and economic capital. *North American Actuarial Journal* **8**, 86–95.

Imakhlaf, A.J., Hou, Y. and Sallak, M. (2017). Evaluation of the reliability of non-coherent systems using binary decision diagrams. IFAC PapersOnLine 50, 12243–12248. https://doi.org/10.1016/j.ifacol.2017.08.2132.

Joe, H. (1997). *Multivariate Models and Dependence Concepts.* Chapman and Hall, London.

Kochar, S., Mukerjee, H. and Samaniego, F.J. (1999). The "signature" of a coherent system and its application to comparison among systems. *Naval Research Logistics* **46**, 507–523.

Krakowski, M. (1973). The relevation transform and a generalization of the Gamma distribution function. *Revue Francaise D'Automatique, Informatique et Recherche Operationnelle* Mai (V-2), 107–120.

Kuo, W. and Zhu, X. (2012). Some recent advances on importance measures in reliability. *IEEE Transactions on reliability* **61**, 344–360.

Li, X., and Ding, W. (2010). Optimal allocation of active redundancies to k-out-of-n systems with heterogeneous components. *Journal of Applied Probability* **47**, 254–263.

Lindqvist, B.H., and Samaniego, F.J. (2019). Some new results on the preservation of the NBUE and NWUE aging classes under the formation of coherent systems. *Naval Research Logistics* **66**, 430–438.

Marichal, J.L., Mathonet, P., Navarro, J. and Paroissin, C. (2017). Joint signature of two or more systems with applications to multi-state systems made up of two-state components. *European Journal of Operational Research* **263**, 559–570.

Marichal, J.-L., Mathonet, P. and Waldhauser, T. (2011). On signature-based expressions of system reliability. *Journal of Multivariate Analysis* **102**, 1410–1416.

Miziuła, P. and Navarro, J. (2018). Bounds for the reliability of coherent systems with heterogeneous components. *Applied Stochastic Models in Business and Industry* **34**, 158–174.

Miziuła, P. and Navarro, J. (2019). Birnbaum importance measure for reliability systems with dependent components. *IEEE Transactions on Reliability* **68**, 439–450.

Müller, A. and Stoyan, D. (2002). *Comparison Methods for Stochastic Models and Risks*. Wiley, New York.

Nakagawa, T. (2008). *Advanced Reliability Models and Maintenance Policies*. Springer Series in Reliability Engineering. Springer-Verlag, London.

Navarro, J. (2016). Stochastic comparisons of generalized mixtures and coherent systems. *Test* **25**, 150–169.

Navarro, J. (2018a). Preservation of DMRL and IMRL aging classes under the formation of order statistics and coherent systems. *Statistics and Probability Letters* **137**, 264–268.

Navarro, J. (2018b). Stochastic comparisons of coherent systems. *Metrika* **81**, 465–482.

Navarro, J. (2018c). Distribution-free comparisons of residual lifetimes of coherent systems based on copula properties. *Statistical Papers* **59**, 781–800.

Navarro, J., Arriaza, A. and Suárez-Llorens, A. (2019). Minimal repair of failed components in coherent systems. *European Journal of Operational Research* **279**, 951–964.

Navarro, J., Balakrishnan, N. and Samaniego, F.J. (2008). Mixture representations of used systems' residual lifetimes. *Journal of Applied Probability* **45**, 1097–1112.

Navarro, J., Belzunce, F. and Ruiz, J.M. (1997). New stochastic orders based on doubly truncation. *Probability in the Engineering and Informational Sciences* **11**, 395–402.

Navarro, J. and Calì, C. (2019). Inactivity times of coherent systems with dependent components under periodical inspection. *Applied Stochastic Models in Business and Industry* **35**, 871–892.

Navarro, J. and del Águila, Y. (2017). Stochastic comparisons of distorted distributions, coherent systems and mixtures with ordered components. *Metrika* **80**, 627–648

Navarro, J., del Águila, Y., Sordo, M.A. and Suárez-Llorens, A. (2013). Stochastic ordering properties for systems with dependent identically distributed components. *Applied Stochastic Models in Business and Industry* **29**, 264–278.

Navarro, J., del Águila, Y., Sordo, M.A. and Suárez-Llorens, A. (2014). Preservation of reliability classes under the formation of coherent systems. *Applied Stochastic Models in Business and Industry* **30**, 444–454.

Navarro, J., del Águila, Y., Sordo, M.A. and Suárez-Llorens, A. (2016). Preservation of stochastic orders under the formation of generalized distorted distributions. Applications to coherent systems. *Methodology and Computing in Applied Probability* **18**, 529–545.

Navarro, J. and Durante F. (2017). Copula-based representations for the reliability of the residual lifetimes of coherent systems with dependent components. *Journal of Multivariate Analysis* **158**, 87–102.

Navarro, J., Durante, F. and Fernández-Sánchez, J. (2021). Connecting copula properties with reliability properties of coherent systems. *Applied Stochastic Models in Business and Industry* **37**, 496–512.

Navarro, J. and Fernández-Sánchez, J. (2020). On the extension of signature-based representations for coherent systems with dependent non-exchangeable components. *Journal of Applied Probability* **57**, 429–440.

Navarro, J. and Fernández-Martínez, P. (2021). Redundancy in systems with heterogeneous dependent components. *European Journal of Operational Research* **290**, 766–778.

Navarro, J., Fernández-Martínez, P., Fernández-Sánchez, J., Arriaza, A. (2020). Relationships between importance measures and redundancy in systems with dependent components. *Probability in the Engineering and Informational Sciences* **34**, 583–604.

Navarro, J. and Gomis, M.C. (2016). Comparisons in the mean residual life order of coherent systems with identically distributed components. *Applied Stochastic Models in Business and Industry* **32**, 33–47.

Navarro, J. and Hernández, P.J. (2004). How to obtain bathtub-shaped failure rate models from normal mixtures. *Probability in the Engineering and Informational Sciences* **18**, 511–531.

Navarro, J. and Hernández, P.J. (2008a). Negative mixtures, order statistics and systems. In: *Advances in Mathematical and Statistical Modelling*. Edited by B.C. Arnold, N. Balakrishnan, J.M. Sarabia, and R. Mínguez, Birkhäuser, Boston, 89–100.

Navarro, J. and Hernández, P.J. (2008b). Mean residual life functions of finite mixtures and systems. *Metrika* **67**, 277–298.

Navarro, J., Pellerey, F. and Di Crescenzo, A. (2015). Orderings of coherent systems with randomized dependent components. *European Journal of Operational Research* **240**, 127–139.

Navarro, J., Pellerey, F. and Longobardi, M. (2017). Comparison results for inactivity times of *k*-out-of-*n* and general coherent systems with dependent components. *Test* **26**, 822–846.

Navarro, J. and Rubio, R. (2010). Computations of signatures of coherent systems with five components. *Communications in Statistics Simulation and Computation* **39**, 68–84.

Navarro, J. and Rubio, R. (2011). A note on necessary and sufficient conditions for ordering properties of coherent systems with exchangeable components. *Naval Research Logistics* **58**, 478–489.

Navarro, J., Ruiz, J.M. and del Águila, Y. (2008). Characterizations and ordering properties based on log-odds functions. *Statistics* **42**, 313–328.

Navarro, J., Ruiz, J. M. and Sandoval, C. J. (2007). Properties of coherent systems with dependent components. *Communications in Statistics Theory and Methods* **36**, 175–191.

Navarro, J. and Rychlik, T. (2007). Reliability and expectation bounds for coherent systems with exchangeable components. *Journal of Multivariate Analysis* **98**, 102–113.

Navarro, J., Samaniego, F.J., Balakrishnan, N. and Bhattacharya, D. (2008). On the application and extension of system signatures to problems in engineering reliability. *Naval Research Logistics* **55**, 313–327.

Navarro, J. and Sarabia, J.M. (2020). Copula representations for the sums of dependent risks: models and comparisons. *Probability in the Engineering and Informational Sciences*. Published online first 2020. https://doi.org/10.1017/S0269964820000649.

Navarro, J. and Shaked, M. (2006). Hazard rate ordering of order statistics and systems. *Journal of Applied Probability* **43**, 391–408.

Navarro, J. and Shaked, M. (2010). Some properties of the minimum and the maximum of random variables with joint logconcave distributions. *Metrika* **3**, 313–317.

Navarro, J. and Spizzichino, F. (2010). Comparisons of series and parallel systems with components sharing the same copula. *Applied Stochastic Models in Business and Industry* **26**, 775–791.

Navarro, J. and Spizzichino, F. (2020). Aggregation and signature based comparisons of multi-state systems via decompositions of fuzzy measures. *Fuzzy Sets and Systems* **396**, 115–137.

Navarro, J., Spizzichino, F. and Balakrishnan, N. (2010). Applications of average and projected systems to the study of coherent systems. *Journal of Multivariate Analysis* **101**, 1471–1482.

Navarro, J., Torrado, N. and del Águila, Y. (2018). Comparisons between largest order statistics from multiple-outlier models with dependence. *Methodology and Computing in Applied Probability* **20**, 411–433.

Nelsen, R.B. (2006). *An Introduction to Copulas*. Springer, New York.

Okolewski, A. (2017). Extremal properties of order statistic distributions for dependent samples with partially known multidimensional marginals. *Journal of Multivariate Analysis* **160**, 1–9.

Parzen, E. (1999). *Stochastic Processes*. Society for Industrial and Applied Mathematics, Philadelphia, PA, USA.

Pellerey, F. and Petakos, K. (2002). On the closure of the NBUC class under the formation of parallel systems. *IEEE Transactions on Reliability* **51**, 452–454.

Ramamurthy, K.G. (1990). *Coherent Structures and Simple Games. Theory and Decision Library.* Series C: Game Theory, Mathematical Programming and Operations Research, 6. Kluwer Academic Publishers Group, Dordrecht.

Rychlik, T., Navarro, J. and Rubio, R. (2018). Effective procedure of verifying stochastic ordering of system lifetimes. *Journal of Applied Probability* **55**, 1261–1271.

Samaniego, F.J. (1985). On closure of the IFR class under formation of coherent systems. *IEEE Transactions on Reliability* **R-34**, 69–72.

Samaniego, F.J. (2007). *System Signatures and Their Applications in Engineering Reliability.* International Series in Operations Research & Management Science, Vol. 110, Springer, New York.

Samaniego, F.J. and Navarro, J. (2016). On comparing coherent systems with heterogeneous components. *Advances in Applied Probability* **48**, 88–111.

Shaked, M. and Shanthikumar, J.G. (1992). Optimal allocation of resources to nodes of parallel and series systems. *Advances in Applied Probability* **24**, 894–914.

Shaked, M. and Shanthikumar, J.G. (2007). *Stochastic Orders*. Springer, New York.

Shaked, M. and Suárez–Llorens, A. (2003). On the comparison of reliability experiments based on the convolution order. *Journal of the American Statistical Association* **98**, 693–702.

Torrado, N., Arriaza, A. and Navarro, J (2021). A study on multi-level redundancy allocation in coherent systems formed by modules. *Reliability Engineering and System Safety* **213**, 107694.

Wang, S. (1996). Premium calculation by transforming the layer premium density. *Astin Bulletin* **26**, 71–92.

Yaari, M.E. (1987). The dual theory of choice under risk. *Econometrica* **55**, 95–115.

Index

© The Editor(s) (if applicable) and The Author(s), under exclusive license   173
to Springer Nature Switzerland AG 2022
J. Navarro, *Introduction to System Reliability Theory*,
https://doi.org/10.1007/978-3-030-86953-3